Lactoferrin

Natural • Multifunctional • Antimicrobial

A.S. Naidu
Director, Center for Antimicrobial Research
Department of Food, Nutrition & Consumer Sciences
California State Polytechnic University
Pomona, California

FCN

CRC Press
Boca Raton London New York Washington, D.C.

Cover Image: Ribbon structure of bovine lactoferrin drawn by RasMol ver.2.6 software (Biomolecular Structures Group, Glaxo R&D, Greenford, UK) using the Brookhaven PDB files (TRFL-BOVIN) from Swiss-Prot database.

Library of Congress Cataloging-in-Publication Data

Catalog record is available from the Library of Congress.

Direct all inquiries to CRC Press LLC, 2000 N.W. Corporate Blvd., Boca Raton, Florida 33431.

International Standard Book Number 0-8493-0909-3
Printed in the United States of America 1 2 3 4 5 6 7 8 9 0
Printed on acid-free paper

Preface

Demands and concerns of the global consumer play a critical role in dictating the momentum of research and development in food protection. The demand for minimally processed foods and sustained functionality of natural bioactive ingredients is steadily rising. This appeal is founded on growing concerns about the use of synthetic preservatives with limited documentation on safety and tolerance; the suspected link between the overuse of antibiotics and the development of multi-drug resistance in microbes; and the increased media dissemination of knowledge on healthy diets. Furthermore, the recent emergence of food-borne pathogens, such as the enterohemorrhagic *Escherichia coli* 0157:H7, multi-drug resistant *Salmonella typhimurium* DT104, and *Enterococcus faecium* has made food safety a high priority with federal, state, and local health and regulatory authorities.

Natural antimicrobials contribute to the host defense of animal and plant species. These compounds exhibit antimicrobial activity when added to foods as natural ingredients. Milk is the first complete functional food devised by nature for protection and development of a newborn mammal. The prophylactic and therapeutic benefits of milk and its bioactive components have been recognized for centuries. The antimicrobial and nutraceutical potential of many components in milk has unfolded in recent years. The dairy industry has begun to isolate these substances from a variety of sources, including cheese whey and hyperimmunized colostrum from cows. Potential applications of such compounds are being developed in the food and health industries.

Lactoferrin (LF) is a metal-binding glycoprotein present in milk and various exocrine secretions that bathe the mucosal surfaces. This natural antimicrobial agent is a multifunctional bioactive molecule with a critical role in many important physiological pathways. LF could elicit a variety of inhibitory effects against microorganisms, comprising stasis, cidal, adhesion–blockade, cationic, synergistic, and opsonic mechanisms. Broad-spectrum activity against different bacteria, viruses, fungi, and parasites, in combination with anti-inflammatory and immunomodulatory properties, make LF a potent innate host defense mechanism.

Recent studies suggest that LF could provide a unique microbial hurdle in the food-processing cascade. LF could inhibit the expression of specific colonizing factors on the microbial surface that promote tissue attachment and compete for tissue-binding sites on the epithelial surface to block or detach microorganisms. Because of this twofold effect, the microorganism is repelled from a biological surface. Since the blocking activity is a molecular phenomenon related to structure–function, the activation of functional sites and a specific target-delivery mechanism are critical for the efficacy of the LF antimicrobial system. This microbial blocking activity of LF has opened a new approach to food safety.

The current global production of bovine milk LF is approximately 100 metric tons and is continuously increasing. This protein is finding applications as an active ingredient in infant formulae and health foods in Southeast Asian countries, in particular. LF is

also in use as a therapeutic and prophylactic agent to control intestinal illnesses and mucosal infections. A number of efficacy studies and clinical trials are ongoing in various laboratories, with over 100 patents filed on this molecule in the past 10 years. Undoubtedly, LF is emerging as one of the leading natural microbial blocking agents in food safety and preservation. This publication is an effort to consolidate the science and technological developments in the fast-growing area of LF research.

This publication is the result of support from special individuals. I am grateful to my colleague, Dr. Wayne Bidlack, Dean, College of Agriculture, California State Polytechnic University, Pomona, for his generous endorsement of my academic endeavors. I am indebted to John Miller, Farmland National Beef, Kansas, for unlocking the opportunities and walking me into the applied world of lactoferrin research. Thanks to the staff at the Life Sciences Division, CRC Press LLC, Boca Raton, Florida, especially, Barbara Norwitz, Lourdes Franco, and Gail Renard for their professionalism and unstinting support in bringing this volume to publication.

A.S. Naidu
January 1, 2000

About the Author

A.S. 'Narain' Naidu is a medical microbiologist with more than 20 years of experience in studying the structure-function relationship of antimicrobial agents. He is considered a leading expert on protective and therapeutic applications of natural antimicrobials. As a public health expert, he has served various agencies, including the World Health Organization, the Hungarian Ministry of Health, the Directorate of Medical and Health Services of the Government of India, and the Royal Swedish Academy of Sciences. Dr. Naidu is currently the Director of the Center for Antimicrobial Research, California State Polytechnic University, Pomona.

Dr. Naidu received his Ph.D. in Microbiology from the Faculty of Medicine, Osmania University, India, in 1985. His research on staphylococcal enterotoxicosis, toxic shock syndrome and milk lactoferrin in Sweden during the mid-1980s has brought him international recognition. He has published over 50 peer-reviewed scientific papers and about 30 book chapters in this area. In 1997, he moved from the University of North Carolina at Chapel Hill to Cal Poly Pomona, to create a new university research center. Dr. Naidu is principal investigator for a number of scientific projects in the areas of food safety, public health, medicine and dentistry. He also supervises a multi-disciplinary team of scientists and research students.

Lactoferrin: Natural•Multifunctional• Antimicrobial

I. INTRODUCTION

Lactoferrin (LF) is an iron-binding glycoprotein present in milk and many exocrine secretions that bathe the mucosal surface. Though the term 'LF' implies an iron-binding component in milk, this molecule co-ordinately binds to various metal ions and occurs in divergent biological milieu including saliva, tears, seminal fluids, mucins, and the secondary granules of neutrophils. LF has a multifunctional role in a variety of physiological pathways and is considered a major component of the preimmune innate defense in mammals. The ability of LF to bind two Fe^{+3} ions with high affinity in cooperation with two HCO_3^- ions is an essential characteristic that contributes to its major structure-functional properties including antimicrobial activities.

Iron is a transition metal belonging to group VII elements in the periodic table; it occurs at an approximate level of 2-g in the body of normal human adults and is essential for all living organisms. The ability of iron to alternate between its two valency states is its most important biological property, and is used in many metabolic pathways including the bioenergetic coupling of inorganic phosphate to adenosine phosphate (ADP) to form adenosine triphosphate (ATP). Nevertheless, this property can be hazardous for the cell, as it can lead to generation of reactive oxygen species. Under controlled conditions, however, iron is used beneficially, e.g., by the phagocytes in intra-lysosomal killing of microorganisms.

Various iron-binding proteins have evolved in the animal physiological system to sequester iron from the milieu. Ovotransferrin (conalbumin) from the egg white was the first iron-binding protein to be purified (Osborne & Campbell, 1900). In 1939, Sörensen and Sörensen identified a red iron-binding protein in bovine milk. Later, Schafer (1951) also reported a similar protein in human milk termed 'siderophilin'. Schade and Caroline (1946) isolated an iron-binding protein from human serum (serotransferrin) which was later named 'transferrin (TF)'. In 1960, Groves from the United Kingdom, Johansson from Sweden, Montreuil and co-workers from France, and Gruttner and co-workers from Germany independently isolated the red milk protein 'lactosiderophilin' from milk.

During the 1960s several investigators isolated and characterized this protein from various exocrine secretions and tissues of humans and animals (Gordon et al., 1963; Blanc et al., 1963; Masson et al., 1965a; 1965b; 1965c; 1966; 1969; Masson & Heremans, 1968). Based on structural and chemical homology with serum TF, Blanc and Isliker (1961a; 1961b) proposed the name 'lactoferrin' for this protein. Though LF is not exclusively found in milk, this name is now widely recognized in the scientific community, although the term 'lactotransferrin' can be found in earlier publications.

LF was isolated as a major component in the specific granules of the polymorphonuclear leukocytes (PMNLs) with an important role in the amplification of inflammatory responses (Masson et al., 1969; Baggiolini et al., 1970; Oseas et al., 1982; Boxer et al., 1982a; 1982b; 1982c; Lash et al., 1983; Ambruso et al., 1984; Brittigan et al., 1989). LF has also been reported to be a component of the sperm-coating antigen (Ashorn et al., 1986) and was found to have cross-reactivity and sequence homology with the major histocompatibility antigen (Aguas et al., 1990). Extensive work by Masson and his Belgian group has established a clear role for LF in cellular immunity and has led to the identification of specific LF-receptors on macrophages, intermediation of endotoxic shock and hyposideremia (Van Snick et al., 1974; Van Snick & Masson, 1976). Pioneering efforts by Spik, Montreuil and their French group unraveled the biological chemistry of LF (Mazurier et al., 1981; Spik et al., 1982; Metz-Boutigue et al., 1984). Lönnerdal has opened the nutritional role for LF in the absorption of metal ions in the intestinal tract (Davidson & Lönnerdal, 1986; Davidson et al., 1990; 1994). In 1978, Broxmeyer and co-workers reported a regulatory function for LF in myelopoiesis (Broxmeyer et al., 1978; 1984).

In 1961, Oram and Reiter reported the ability of milk LF to inhibit the growth (stasis effect) of *Bacillus* sp. and found that nutritional deprivation of the bacteria from iron accounted for the antimicrobial activity. The antimicrobial spectrum of LF was further elucidated by Brock's group (Brock et al., 1978; 1983; Brines & Brock, 1983). In 1977, Arnold and co-workers reported cidal activity for LF against a variety of microorganisms at acidic pH and noted that it is not inhibited by salts in the media (Arnold et al., 1977; 1980; 1981). Despite wide citation in literature, the cidal effect of native LF appears to be artifactual in nature, caused by reactants other than LF in the milieu (Lassiter, 1990). The research group at the Morinaga Milk Company in Japan found that acid/pepsin hydrolysis of bLF could generate cationic antimicrobial peptides 'lactoferricins (LFcins)' (Tomita et al., 1989; 1991; 1994; Bellamy et al., 1992; 1993). This broad-spectrum cationic microbial killing seems to be non-specific and readily inhibited by salts in the milieu at physiological concentrations. Erroneously, LFcins are widely described in the literature as bactericidal domains, implying they are peptides responsible for the antimicrobial effect reported by Arnold's group.

Interaction of LF with specific targets on the microbial surface causes an array of outcomes either to the advantage of the host (microbial blocking effects) or the microorganism (iron-acquisition and pathogenesis).

Naidu and co-workers have identified, isolated and characterized LF-binding microbial targets in a variety of Gram-positive and Gram-negative bacteria (Naidu et al., 1990; 1991a; 1991b; 1992; 1993). Specific high-affinity interaction of LF with pore-forming outer membrane proteins (OMPS) of *Escherichia coli*, in particular, has unraveled a molecular mechanism for antimicrobial activity which seems to be well conserved in

Gram-negative enterics (Kishore et al., 1991; Tigyi et al., 1992; Erdei et al., 1993; Naidu & Arnold, 1994). The LF-mediated outer membrane damage in Gram-negative bacteria reported by Ellison et al. (1988) has explained certain antimicrobial effects such as antibiotic potentiation, release of lipopolysaccharides (LPS) and alterations in microbial OM permeation.

The critical role of iron in the pathogenesis of many microbial infections has been widely advocated (see reviews: Weinberg et al., 1975; Bullen, 1981). Thus, the mobilization of iron from physiological milieu from LF, TF and ferritin by pathogens appeared to be an important virulence trait. Rogers and Synge (1978) reported a siderophore-mediated mobilization of iron from LF by *E.coli*. Another type of iron-acquisition mechanism involving specific receptors was reported by Alderete's group as well as Sparling and co-workers, as a virulence factor in various intracellular pathogens, in particular, among the etiological agents of sexually transmitted diseases (Mickelsen et al., 1982; Peterson & Alderete, 1984; Biswas & Sparling, 1995). Later, Schryvers' group identified and characterized a number of specific receptors for LF on various mucosal pathogens (Schryvers & Morris, 1988; Schryver, 1989).

This chapter is mainly confined to the antimicrobial spectrum of milk LF and its possible role in food safety and preservation. The multifunctionality of LF is also described to emphasize its nutraceutical benefits as a value-added food ingredient.

II. OCCURRENCE

A. Normal levels

In the human body, LF occurs in two major reservoirs, a circulatory pool stored in the polymorphonuclear lymphocytes (PMNL) and a stationary pool on the mucosal surfaces. In PMNL, LF is associated with the secondary (specific) granules at a concentration of about 15 $\mu g/10^6$ cells (Baggiolini et al., 1970; Bennett & Kokocinski, 1978) and is released isochromously with other lysosomal proteins into the plasma during phagocytosis (Bennett & Skosey, 1977; Leffell & Spitznagel, 1975). LF content in plasma is low or undetectable during agranulocytosis (Bennett & Kokocinski, 1978) and neutropenia (Hansen et al., 1975; Olofson et al., 1977), thus suggesting that PMNL are the sole source of intravascular LF production. The concentration of LF in human plasma is about 0.2 to 1.5 μg/ml (Rumke et al., 1971); the values are comparatively lower in women than in men (Bennett & Mohla, 1976). The plasma LF levels are also low in children due to reduced PMNL secretion (Gahr et al., 1987). LF is rapidly eliminated from the plasma with a mean fractional catabolic rate of 5.7/day, by liver and spleen; however, apo-LF is removed at a slower rate of 1.22/day (Bennett & Kokocinski, 1979).

Hirai and co-workers (1990) measured LF concentration in human milk and colostrum from 1 to 60 days after parturition (125 samples) by rocket immuno-electrophoretic assay using anti-hLF antiserum. The LF concentrations in colostrum (1-3 days of puerperium, n = 35), the transitional milk (4-7 days, n = 60), and mature milk (20-60 days, n = 30) were about 6.7, 3.7, and 2.6 g/L, respectively. Both the LF and total protein (TP) concentrations showed inverse correlation with the days after parturition. The LF/TP ratio in the mature milk (16.1%) was significantly less than that in the colostrum (20.4 %) and the transitional milk (21.4%). Furthermore, iron concentration in human milk was also measured by the internal standard technique of the spiked method on atom-

ic absorption, and the LF iron-saturation was calculated. Neither Fe nor iron-saturation % showed a significant difference among these three stages of lactation. The means (n = 125) of Fe and iron-saturation % were about 0.61 μg/ml and 11.8%, respectively. However, significant correlation was observed between LF and Fe or between LF/TP and both Fe and TP in the mature milk. These results suggest that the mechanism stimulating the synthesis and secretion of LF is different from those of other proteins and LF can play variable roles in iron nutrition of infants during different stages of lactation.

An immunoperoxidase staining technique was used for detecting LF, TF and ferritin in routine histological paraffin sections of human tissue. LF was found in lactating breast tissue, bronchial glands, PMNLs, and gastric and duodenal epithelial cells (Mason & Taylor, 1978). The expression of LF was studied in human gastric tissues displaying normal, benign hyperplastic or malignant histology (Luqmani et al., 1991). A single 2.5-kb mRNA was detected in only 14% (2/14) of normal resections. This was similar to the finding that 85% of tumors were also negative, with 4/27 positive. In contrast, samples with superficial or atrophic gastritis had a high frequency of expression, with 5/7 and 9/14 positive, respectively. The higher incidence of LF mRNA in antral samples was a reflection of the greater proportion of these compared with body resections of patients with gastritis. No expression was seen in any of 5 gastric carcinoma cell lines. High levels were observed in the cardia, in contrast to complete absence in the oesophagus. Immunocytochemistry showed localization of LF in cells of both antral and body glands. Chief cells, but not adjacent parietal cells, were strongly stained. In tissues exhibiting superficial or atrophic gastritis a greater degree and intensity of staining was observed as compared with samples with normal histology

LF accounts for about 11.5% of the total secretary proteins of the bronchial glands (Harbitz et al., 1984). In these glands, LF is uniformly distributed in large amounts in the serous cells and is restricted only to the basal part of cytoplasm in the mucus cells (Masson et al., 1966). LF is released into the nasal secretions by serous cells of submucosal glands under the influence of the parasympathetic nervous system (Raphael et al., 1989). In human tears, LF is synthesized in the lacrimal glands (Janssen & van Bijsterveld, 1982; 1983) at a concentration of 1 to 3 mg/ml that accounts for ~25% of the total tear protein (Kijlstra et al., 1983). In the male reproductive tract, LF is found in the prostate, seminal vesicles, and seminal plasma (0.2 to 1.0 mg/ml), but not in testis (Rumke, 1974; Wichmann et al., 1989). As the sperm-coating antigen, LF may suppress lymphocyte response against sperm (Ashorn et al., 1986). It may also partially account for the antimicrobial and immune-suppressive activity of the seminal fluid in the female reproductive tract before and during fertilization (Broer et al., 1977). In the female reproductive tract, LF has been detected in the cervical mucus and endometrium of the secretory uterus (Tourville et al., 1970)

B. Clinical levels

Acute phase host responses such as an inflammation or toxic shock result in the depletion of iron from plasma (Klasing, 1984). The interleukin 1 (IL-1) seems to mediate multiple aspects of acute phase response and also induce exocytosis of PMNLs to release intracellular granule contents, including LF (Klempner et al., 1978; Dinarello, 1984). Eventually, the plasma LF levels increase in various pathological conditions and its estimation may serve as a prognostic marker. Thus, the elevated plasma LF level is an early

indicator of septicemia and endotoxemia (Gutteberg et al., 1988). During meningococcal septicemia, LF level in the serum and cerebrospinal fluid is elevated within 18 h of onset as to the same magnitude of the C-reactive protein (Gutteberg et al., 1984). Similarly, an increase in the plasma and serum LF levels during cystic fibrosis seems to correlate with the intensity of the cystic fibrosis protein (Barthe et al., 1989). The estimation of elevated LF concentration with cholecytokinnin secretin (CCK-S) test has been suggested to improve the differential diagnosis of chronic pancreatitis (Dite et al., 1989). LF to lysozyme ratio in the crevicular fluid may be helpful in the diagnosis of localized juvenile periodontitis (Friedman et al., 1983).

In the pathogenesis of rheumatoid arthritis, iron has an important role in the degradation of intact cartilage matrices, due to its capacity to generate free radicals that activate latent collagenases (Blake et al., 1981; 1984; Bukhhardt & Schwingel, 1986). Elevated levels of LF in the synovial fluids and synovial membranes due to inflammatory tissue damage is a characteristic of this disease (Ahmadzadeh et al., 1989). This high LF concentration has been suggested in the regulation of subsequent inflammatory processes critical for articular damage.

Neoplastic cells have an increased iron-requirement for the initiation and maintenance of DNA synthesis and for cell multiplication (Gatter et al., 1983; Barresi & Tuccari, 1987). In different types of malignancies, iron uptake may be mediated by specific receptors for LF or TF. Immunohistochemical studies have demonstrated LF in adenocarcinomas of the parotid gland (Caselitz et al., 1981), well differentiated prostatic carcinomas, breast carcinomas (Charpin et al., 1985), thyroid carcinomas of follicular origin (Tuccari & Barresi, 1985; Barresi & Tuccari, 1987), renal cell carcinomas (Loughlin et al, 1987), intestinal-type carcinomas, and incomplete intestinal metaplasia (Tuccari et al., 1989).

III. ISOLATION AND PURIFICATION

LF from bovine milk was first isolated by Groves (1960). Various procedures for isolation of LF from mammalian milk have been reported. The most commonly used methods include chromatographic separation on CM-Sephadex, Cibacron Blue F3G-A-Sepharose, Heparin agarose and single stranded DNA agarose (*FIGURE 1*).

A. Size-exclusion chromatography on CM-Sephadex columns

Gel filtration is one of the earlier techniques used for isolation of LF from various biological secretions (Butler, 1973). Tsuji and co-workers (1989) described a method to isolate LF from bovine colostrum. Colostrum obtained within 24 h after parturition was centrifuged at 3000 rpm for 10 min and 1 liter of the skimmed colostrum was dialyzed against distilled water in the presence of 0.01% sodium azide for 3 days at 4°C with several changes of distilled water. The dialyzed skimmed colostrum was loaded on a CM-Sephadex C-50 (Pharmacia) column (2 x 20 cm) equilibrated with 0.05 M potassium phosphate buffer, pH 8.0 (buffer-A). LF was eluted with a linear gradient of NaCl (0.1 to 0.7 M) in buffer-A after unbound proteins and weakly bound proteins were washed out sequentially with 100 ml of buffer-A and 100 ml of buffer-A containing 0.05 M NaCl. The combined fraction containing LF was dialyzed against buffer-A at 4°C overnight and reloaded onto a column of CM-Sephadex C-50 (1 x 5 cm). LF was eluted with a linear gradient of NaCl (0.2 to 0.5 M) in buffer-A, and at this step, 192 mg of nearly homogenous LF was obtained.

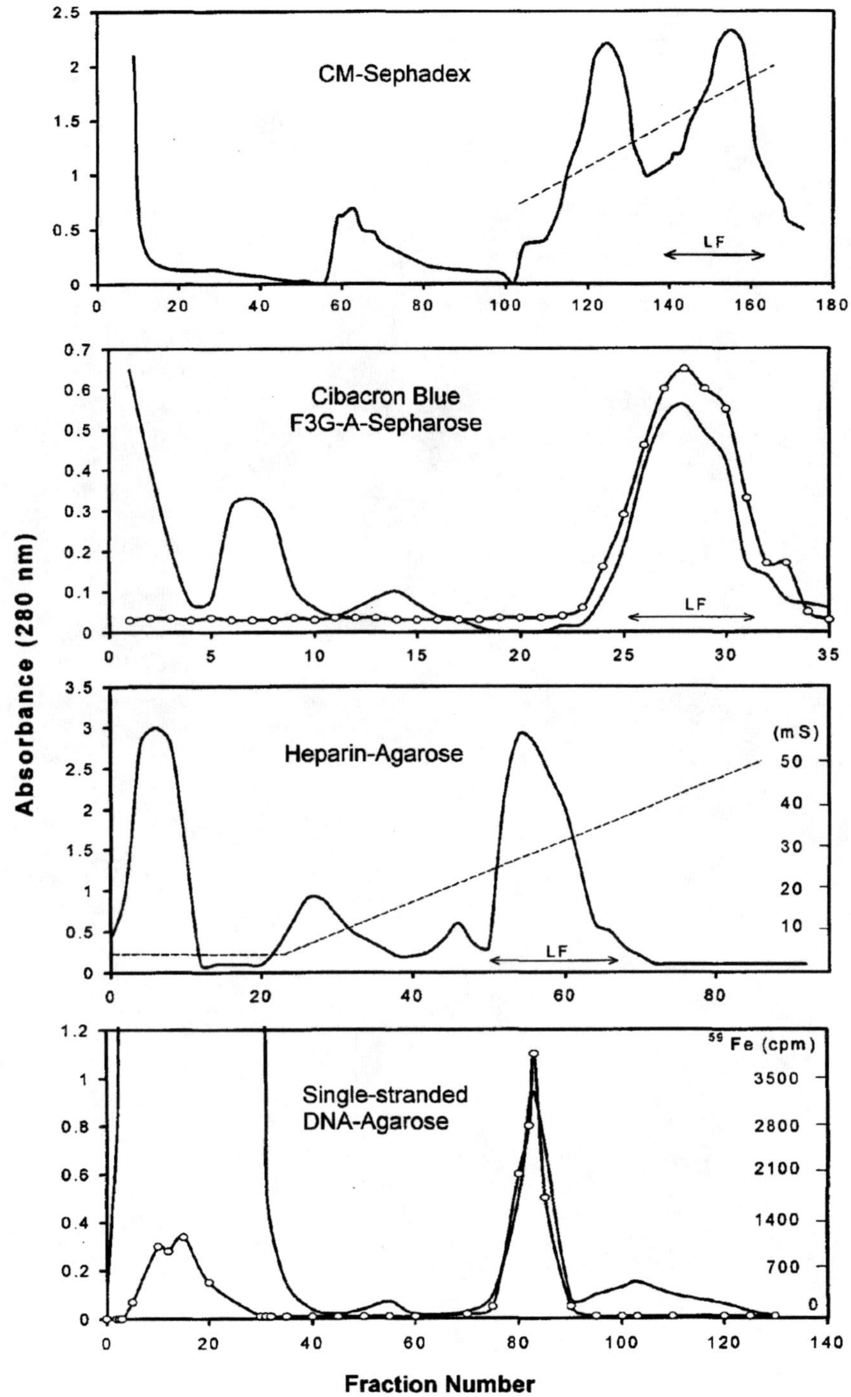

FIGURE 1. LF isolation methods - elution profiles on different chromatography columns [redrawn from Tsuji et al., 1989; Furmanski et al., 1989; Rejman et al., 1989; Hutchens et al., 1989].

B. Affinity-chromatography on Cibachron Blue-Sepharose columns

Various laboratories have adapted this method to isolate LF from human milk (Arnold et al., 1977; Bezwoda & Mansoor, 1986; Furmanski et al., 1989; Shimazaki & Nishio, 1991). Briefly, 5 ml of pooled human milk is skimmed and decaseinated by centrifugation (10,000 g, 40 min), acidification (to pH 4.7 with HCl), heating (40°C, 30 min) and recentrifugation (10,000 g, 40 min). The milk whey is dialyzed overnight against 500 ml veronal buffer. The whey was diluted to 25 ml with veronal buffer. The sample was applied to Cibachron Blue Sepharose CL-6B (Pharmacia) column (1 x 10 cm). The column is washed with 25-ml veronal buffer and eluted with a 30-ml linear gradient of 0.5-1 M NaCl in veronal buffer. A single protein peak corresponding to LF elutes at about 0.75 M NaCl. The radioisotope labeled ^{59}Fe-saturated LF demonstrated an elution pattern similar to apo-LF during this separation method (Furmanski et al., 1989).

C. Affinity-chromatography on heparin-cross linked columns

Heparin-agarose affinity chromatography has been used to isolate LF from human milk whey in a single chromatographic step (Blackberg & Hernell, 1980). Al-Mashikhi and Nakai (1987) have used heparin-sepharose affinity chromatography to isolate LF from cheddar cheese whey. In this procedure, whey is dialyzed against 0.05M NaCl in 0.005 M sodium barbital-HCl buffer, pH 7.4. Whey solution is applied to a heparin-agarose column equilibrated with the above dialysis buffer. Protein is eluted at a flow rate of 48 ml/h using a continuous gradient of 0.05M to 1.0 M NaCl constituted in the dialysis buffer. Fractions are collected and absorbance is read at 280 nm. Rejman et al. (1989) used this method to isolate bLF from mammary secretions collected during the nonlactating period. About 1600 absorbance units (280 nm) of whey protein were efficiently separated by the heparin-agarose column (packed with 2.0 x 16.5 cm of Affi-Gel heparin agarose from Bio-Rad) into four absorbance peaks. LF was identified in the fourth peak (eluting at a conductivity of 30 mS). Both iron-saturated and apo-forms of LF demonstrated similar elution profiles in this technique. IgG1 and secretory component are the trace contaminants with bovine milk whey fractionation. In contrast, serum albumin was identified as the contaminant with human milk whey separation on the heparin column.

D. Affinity-chromatography on DNA-agarose columns

Hutchens and co-workers (1989) reported that immobilized DNA is effective for a rapid and complete purification of apo-LF and holo-LF from colostrum in a single step. Urea was utilized as a mobile phase modifier to eliminate the interaction of other proteins such as serum albumin both with LF and with the immobilized DNA. Briefly, in this method, single-stranded DNA-agarose is packed into a 1.0 x 1.5 cm column to a bed volume of 5 to 10 ml. The column is washed with water, then equilibrated with 20 mM Hepes buffer, pH 8.0, with or without 6 M urea, at a flow rate of 30 ml/h. The separation procedure is performed at room temperature. Solid urea (up to 6 M) is added before the sample (5 to 10 ml) is applied to the DNA affinity column. The column is washed with equilibration buffer and prior to gradient elution, urea (if present) is removed with several column volumes of 20 mM Hepes buffer, pH 8.0 (HB). LF is eluted with a linear gradient of NaCl (0 to 1.0 M) in 20 mM HB. Fractions of 1 ml each are collected and the absorbance

was measured at 280 nm. The radioisotope labeled ^{59}Fe-saturated LF demonstrated an elution pattern similar to apo-LF during this separation method. Finally, after each purification procedure the DNA-agarose column was washed extensively, first with 2 M NaCl in HB, followed by 8 M guanidine-HCl in HB and finally with water.

Iron saturation and desaturation: Various methods of iron saturation (holo-form) and desaturation (apo-form) of LF have been described. Briefly, apo-LF is prepared by dialysis against an acetic acid/sodium acetate buffer (pH 4.0), followed by exhaustive dialysis against deionized distilled water (Mazurier & Spik, 1980). Holo-LF is prepared by adding a large excess of citric acid (pH 2.5; 60 mol citric acid:1 mol iron). After incubation for 10 min, the pH is raised to 7.0 with 0.1 M NaOH. Excess sodium bicarbonate is added (2 mol bicarbonate:1 mol iron). Unbound iron is removed by gel filtration (Azari & Baugh, 1967).

Masson and Heremans (1968) described a method to prepare apo-LF. A concentration of 1% LF solution was deprived of iron by dialysis against 20 volumes of 0.1 M citric acid. After 36 h the citrate was eliminated by dialysis against 20 volumes of deionized water for 2-h at 4°C. The required amount of solid disodium phosphate was added to the dialysis flask. Stirring was discontinued and the flask was kept at 4°C. The pH of the suspension increased slowly as the crystals of disodium phosphate dissolved into the solution. This precaution is needed to prevent precipitation of the protein. Stirring is resumed as soon as the pH in the upper layer of buffer has reached a value of 5.0. The final pH is 7.6. After such treatment, the solution of LF turns completely colorless.

Fe(III) removal from LF by an Fe(III)-chelating resin with immobilized 3-hydroxy-2-methyl-4(1H)-pyridinone ligands at physiological pH in the presence of citrate has been described (Feng et al., 1995). The resin had a marked effect on the extent of iron removal. By using the Fe(III)-chelating resin, removal of iron from LF was nearly complete in < 24 h. Apo-LF with 4% iron saturation could be prepared conveniently from 100% or from 18% iron-saturated LF under mild conditions without affecting the iron-binding capacity of the protein.

IV. MOLECULAR PROPERTIES

A. Physico-chemistry

The physico-chemical characteristics of hLF and bLF are listed in *TABLE 1*. Like the transferrins of blood serum and egg white, LF is a single polypeptide chain with a molecular weight in the range 75 to 80-kDa. Dry weight determinations together with measurement of iron-binding capacities, showed combining weights per iron atom bound of 39,000 for bLF, and 40,000 for hLF (Aisen & Leibman, 1972). Accordingly, these correspond to molecular weights of 78,000 and 80,000 daltons, respectively, for protein molecules with two specific binding sites.

The isoelectric point (pI value) of bLF is reported at about 8.0 by free boundary electrophoretic methods (Groves, 1960; Szuchet-Drechin & Johanson, 1965); pI of 8.8 by Rotafors method (Shimazaki et al., 1993) and a pI of 8.2-8.9 by chromatofocusing (Shimazaki et al., 1993). On the otherhand, a wide range of pI values, from 5.5 to 10.0 have been reported for hLF by isoelectric focusing techniques [5.8 to 6.5 by Bezwoda & Mansoor, 1989; 6.9 by Malmquist & Johanson, 1971; 8.7 by Moguilevsky et al., 1985; 8.8

TABLE 1. Lactoferrin – Physico-chemical properties

Property	Human LF	Bovine LF	Reference
Molecular mass			
Sedimentation co-efficient	75,100	77,200 ± 1,300	Castellino et al., 1970
SDS-PAGE	76,800 ± 1,600	76,000 ± 2,400	Querinjean et al., 1971
Iron titration	80,000	78,500	Aisen & Leibman, 1972
Isoelectric point			Bezwoda & Mansoor, 1989
Chromato focusing	6.8 – 8.0	8.2 – 8.9	Yoshida & Xiuyun, 1991
Isoelectric focusing	5.8 – 6.5	9.5 – 10.0	Shimazaki et al., 1993
Absorption spectra			Aisen & Leibman, 1972
Apo-form at 280 nm	10.9	12.7	
Holo-form at 470 nm	0.510	0.400	
Glycosylation	Relatively high	Low	Metz-Boutigue et al., 1984
Protease sensitivity	Relatively low	High	Brines & Brock, 1983
IgA-complexes	Present	Absent	Watanabe et al., 1984
Iron-binding			
Equilibrium dialysis (K_1 x 10^{-4})	26.0	3.73	Aisen & Leibman, 1972
Thermal denaturation			Paulsson et al., 1993
Apo-LF denaturation (Tmax: °C)		71 ± .3 & 90 ± .3	
Apo-LF enthalpy (ΔH_{cal}: J/g)		12 ± .4 & 2 ± .5	
Holo-LF denaturation (Tmax: °C)		65 ± .3 & 93 ± .3	
Holo-LF enthalpy (ΔH_{cal}: J/g)		2 ± 1 & 37 ± 1	

to 8.9 by Birgens & Kristensen, 1990; and 8-10 by Kinkade et al., 1976]. By the Rotafors method hLF was focused at a pI of 8.7 and by chromatofocusing the hLF was eluted at pH 6.8-8.0 (Shimazaki et al., 1993).

Heat-induced enthalpy changes in different forms of bLF in water were examined by differential scanning calorimetry (Paulsson et al., 1993). Two thermal transitions with varying enthalpies were observed, depending on the iron-binding status of the protein. Iron-saturated holo-LF was more resistant to heat induced changes than the apo-form.

Investigations of metal-substituted hLF by fluorescence, resonance Raman, and electron paramagnetic resonance (EPR) spectroscopy confirm the close similarity between LF and serum TF (Ainscough et al., 1980). As in the case of Fe(III)- and Cu(II)-TF, a significant quenching of apo-LF's intrinsic fluorescence is caused by the interaction of Fe(III), Cu(II), Cr(III), Mn(III), and Co(III) with specific metal binding sites. Laser excitation of these metal-LFs produce resonance Raman spectral features at about 1605, 1505, 1275, and 1175 cm-1. These bands are characteristic of tyrosinate to the metal ions, as has been observed previously for serum TF, and permit the principal absorption band (l_{max} between 400 and 465 nm) in each of the metal-LFs to be assigned to charge transfer between the metal ion and tyrosinate ligands. Furthermore, as in serum TF the two metal binding sites in LF can be distinguished by EPR spectroscopy, particularly with the Cr(III)-substituted protein. Only one of the two sites in LF allows displacement of Cr(III) by Fe(III). LF is known to differ from serum TF in its enhanced affinity for iron. Accordingly, the kinetic studies show that the rate of uptake of Fe(III) from Fe(III)-citrate is 10 times faster for apo-LF than for apo-TF. Furthermore, the more pronounced conformational change which occurs upon metal binding to LF is corroborated by the production of additional EPR-detectable Cu(II) binding sites in Mn(III)-LF. The lower pH required for iron removal from LF causes some permanent change in the protein as judged

by altered rates of Fe(III) uptake and altered EPR spectra in the presence of Cu(II). Thus, the common method of producing apo-LF by extensive dialysis against citric acid (pH 2) appears to have an adverse effect on the protein.

The anion binding properties of hLF, with Fe(III) or Cu(II) as the associated metal ion, highlight differences between the two sites, and in the anion binding behavior when different metals are bound (Brodie et al., 1994). Carbonate, oxalate and hybrid carbonate-oxalate complexes have been prepared and their characteristic electronic and EPR spectra recorded. Oxalate can displace carbonate from either one or both anion sites of $Cu_2(CO_3)_2$ LF, depending on the oxalate concentration, but no such displacement occurs for $Fe_2(CO_3)_2$ LF. Addition of oxalate and the appropriate metal ion to apo-LF under carbonate-free conditions gives dioxalate complexes with both Fe^{3+} and Cu^{2+}. Both the carbonate and oxalate ions bind in bi-dentate fashion to the metal, except that the carbonate ion in the N-lobe site of dicupric LF is mono-dentate. The hybrid copper LF complex shows that the oxalate ion binds preferentially in the C-lobe site in a bi-dentate mode. Overall these observations lead to a generalized model for synergistic anion binding by TF and allow comparisons to be made with non-synergistic anions such as citrate and succinate.

The amino acid composition of LF molecules isolated from the milk of different mammalian species is shown in *TABLE 2*. Tryptic peptide maps of hLF show some 40 spots, which is a much smaller number than would be predicted from the lysine and arginine content (Querinjean et al., 1971). LF is a glycoprotein containing two glycans attached through *N*-glycosidic linkages. The two *N*-acetyl-lactosaminic-type glycans are structurally heterogenous (Spik et al., 1982) and differ from those of other transferrins (Spik et al., 1975; Dorland et al., 1979; Van Halbeck et al., 1981). Analysis of bLF for carbohydrate content reveals 1 residue of terminal sialic acid, 10 to 11 residues of *N*-

TABLE 2. Lactoferrin - Amino acid composition of protein isolated from different species

Amino acid	Human	Bovine	Porcine	Monkey	Murine	Equine
Aspartate	71	71	57	72	71	72
Threonine	31	39	28	31	40	32
Serine	50	45	46	43	50	48
Glutamate	70	73	71	67	71	70
Proline	35	31	30	34	33	33
Glycine	56	43	50	61	51	50
Alanine	63	59	56	73	60	71
Cysteine	32	28	X	23	28	38
Methionine	6	4	4	2	5	3
Valine	49	43	42	50	45	44
Isoleucine	16	17	18	15	17	14
Leucine	61	61	53	62	57	60
Tyrosine	20	19	20	29	18	20
Phenylalanine	31	25	26	30	25	25
Tryptophan	11	9	X	X	10	12
Lysine	46	42	40	47	55	45
Histidine	9	10	8	9	10	10
Arginine	46	32	36	40	32	33

Adapted from Hutchens et al., 1989

acetyl glucosamine, 5 to 6 residues of galactose, and 15 to 16 residues of mannose per molecule (Castellino et al., 1970). In hLF, according to the sequence studies by Metz-Boutigue et al., 1984), asparagine residues 137 and 490 are glycosylated. Prediction of the secondary structure suggested that the two prosthetic sugar groups were linked to asparagine residues located in a β-turn conformation. The non-glycosylated asparagine residue 635 also occurs in a β-turn whereas asparagine residue 389 is located in a region of non-predictable structure. Spik et al. (1982) elucidated the primary structure of glycans from hLF. The polypeptide chain of hLF consists of two glycoslyation sites to which glycans are linked through an *N*-(β-aspartyl)-*N*-acetylglucosaminylamine bond and which are structurally heterogenous. After chymotryptic or pronase digestions, glycopeptides with five different glycan structures were isolated. Three of these structures were determined by using methanolysis, methylation analysis, hydrazinolysis/nitrous deamination/ enzymatic cleavage and 1H-NMR spectroscopy at 360 MHz. *Glycopeptides-A/B:* NeuAc(α-2-6)Gal(β-1-4)GlcNAc(β-1-2)Man(α-1-3)[NeuAc(α-2-6)Gal(β-1-4)GlcNAc(β-1-2)Man(α-1-6)]Man(β-1-4)GlcNAc(β-1-4)[Fuc(α-1-6)]GlcNAc(β-1-)Asn;*Glyco-peptide-C:*NeuAc(α-2-6)(Gal(β-1-4)GlcNAc(β-1-2)Man(α-1-3)(Gal(β-1-4)[Fuc(α-1-3)] GlcNAc(β-1-2)Man(α-1-6))Man(β-1-4)GlcNAc(β-1-4)[Fuc(α-1-6)]GlcNAc(β-1-)Asn. *Glycopeptide-D:*NeuAc(α-2-6)Gal(β-1-4)GlcNAc(β-1-2)Man(α-1-3)[Gal(β-1-4)GlcNAc (β-1-2)Man(α-1-6)]Man(β-1-4)Glc-NAc(β-1-4)[Fuc(α-1-6)]GlcNAc(β-1)Asn. The other two glycopeptides were obtained in very low amounts with more complex structures.

It is generally believed that each iron-binding site contains two or three tyrosine residues (Windle et al., 1963) and one or two histidine residues (Krysteva et al., 1975; Mazurier et al., 1981); and concomitantly bound bicarbonate ion (Schlabach & Bates, 1975) may be held electrostatically to an arginyl side group (Rogers et al., 1978).

B. Structure

LF is a member of the iron-binding protein family collectively known as transferrins (TF). Human LF demonstrates amino acid sequence homology (more pronounced in the C-terminal region) with serum hTF (59%) and hen ovotransferrin (49%). Computer analysis has established an internal homology of the two lobes (residues 1-338, and 339-703), each containing a glycosylation site (aspargine residues 137 and 490) located in homologous position (Metz-Boutigue et al., 1984). Each lobe has a capacity to bind one Fe^{+3} ion with high affinity ($K_d = 10^{-20}$ M^{-1}) in the presence of a carbonate or bicarbonate anion (Harris, 1986). It has been suggested that the iron-binding site contain two or three tyrosine residues (Windle et al., 1963; Teuwissen et al., 1972) and one or two histidine residues (Mazurier et al., 1981); the concomitantly bound bicarbonate anion (Schlabach & Bates, 1975) may be held electrostatically to an arginyl side group (Rogers et al., 1978).

Baker and co-workers have extensively studied the three-dimensional structure of LF (*FIGURE 2*).The hLF molecule at 3.2-Å resolution has two-fold internal homology. The N- and C-terminal halves form two separate globular lobes, connected by a short alpha-helix, and carry one iron-binding site each (Anderson et al., 1987). The two lobes of the molecule have very similar folding; the only major differences being in surface loops. Each lobe is subdivided into two dissimilar α/β domains, one based on a six-stranded mixed β-sheet, the other on a five-stranded mixed β-sheet, with the iron site in the inter-domain cleft. The two iron sites appear identical at the present resolution. Each iron

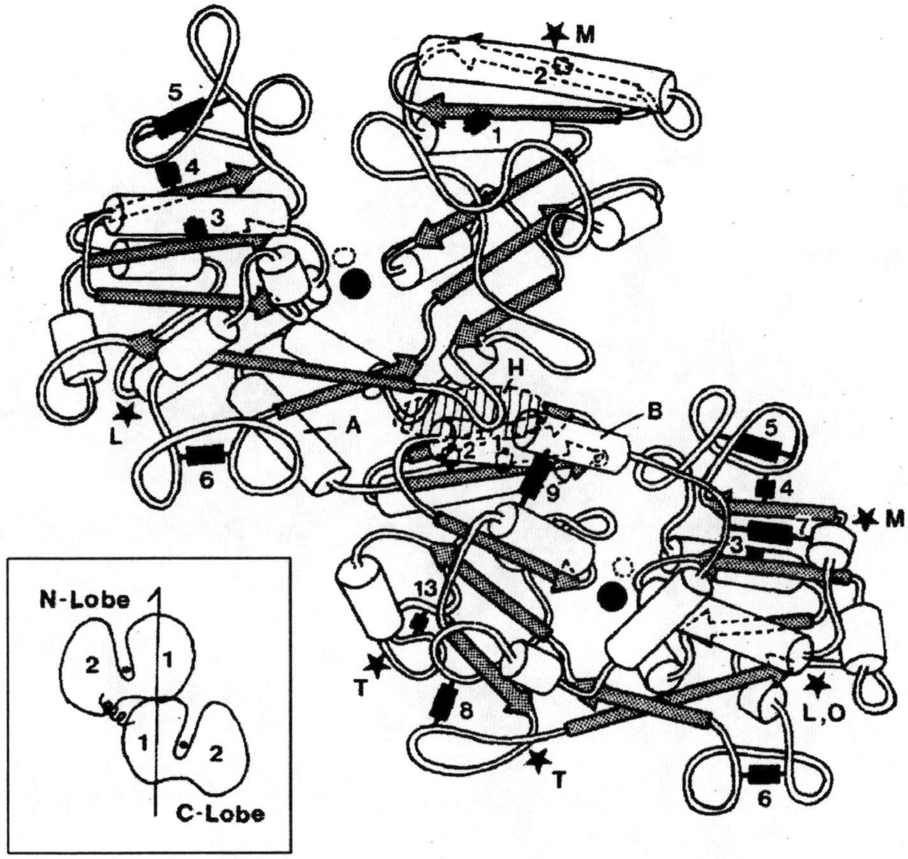

FIGURE 2. LF molecule: A schematic diagram. Helices are shown as cylinders, β-strands as arrows, iron atoms ●, probable anion sites ○, disulfide bridges ■, and carbohydrate attachment sites ★ (labeled L for lactoferrin, T for serum transferrin, O for ovotransferrin, and M for melanotransferrin). The N-terminal half (N-lobe) is at top, the C-terminal half (C-lobe) at bottom; their relative orientations, related by a two-fold screw axis are shown in the inset. The two domains in each lobe are labeled 1 and 2. A region where hydrophobic interactions between the two lobes are made is indicated (H). Helices labeled are the connecting helix (A) and the C-terminal helix (B) [from Baker et al. (1987) with permission from the Elsevier Publications].

atom is coordinated to four protein ligands, 2 tyrosine, 1 aspartate, 1 histidine, and the specific CO_3^{2-}, which appears to bind to iron in a bi-dentate mode. The anion occupies a pocket between the iron and two positively charged groups on the protein, an arginine side-chain and the N terminus of helix 5, and may serve to neutralize this positive charge prior to iron binding. A large internal cavity, beyond the arginine side-chain, may account for the binding of larger anions as substitutes for CO_3^{2-}. Residues on the other side of the iron site, near the inter-domain crossover strands, could provide secondary anion binding sites, and may explain the greater acid-stability of iron binding by LF, compared with serum TF (Anderson et al., 1989).

X-ray structure analyses of four different forms of hLF (diferric, dicupric, an oxalate-substituted dicupric, and apo-LF), and of bovine diferric LF, have revealed various ways in which the protein structure adapts to different structural and functional states (Baker et al., 1991). Comparison of diferric and dicupric LFs revealed that different metals, through slight variations in the position, have different stereo-chemistry and anion coordination without any significant change in the protein structure. Substitution of oxalate for carbonate indicated that small side-chain movements in the binding site could accommodate larger anions. The multi-domain nature of LF also allows rigid body movements. The structure of apo-LF demonstrated the importance of large-scale domain movements for metal binding and release and suggested equilibrium between open and closed forms in solution, with the open form being the active binding species.

The crystal structure of a site-specific mutant of the N-terminal half-molecule of hLF(N), in which the iron ligand aspartate-60 has been mutated to serine, was studied to determine the effects of the mutation on iron binding and domain closure (Faber et al., 1996). At the mutation site the serine side-chain neither bound to the iron atom nor made any inter-domain contact similar to the substituted aspartate; instead a water molecule filled the iron coordination site and participated in inter-domain hydrogen bonding. The domain closure was also changed, with the mutant with a more closed conformation. Consideration of crystal packing suggested that the altered domain closure is a genuine molecular property but both the iron coordination and inter-domain contacts were consistent with weakened iron binding in the mutant.

The role of conserved histidine ligand in iron binding of LF was studied by site-directed mutagenesis and X-ray crystallographic analysis (Nicholson et al., 1997). *His-253* in the N-terminal half-molecule of hLF (residues 1-333) was changed to *Gly, Ala, Pro, Thr, Leu, Phe, Met, Tyr, Glu, Gln,* and *Cys* by oligonucleotide-directed mutagenesis. The mutant proteins were expressed in baby hamster kidney cells, at high levels, and purified. The study indicated that the *His* ligand is essential for the stability of the iron binding site. All of the substitutions destabilized iron binding irrespective of whether the replacements were potential iron ligands or not. Iron was lost below pH approximately 6.0 for the *Cys, Glu,* and *Tyr* mutants and below pH 7.0 or higher for the others, compared with pH 5.0 for the N-terminal half molecule. The destabilization was attributed to both steric and electronic effects. The decreased stability of the iron binding was attributed solely to the loss of the *His* ligand as the protein conformation and inter-domain interactions were unchanged.

		1	10		20		30
Human	G	RRRRS	VQWCA	VSQPE	ATKCF	QWQRN	MRKVR
Bovine		APRKN	VRWXT	ISQPE			
Porcine		APKKG	VRWCV	ISTAE	YSKCR	QWQSK	I RRTN
Murine		KATT	VRWCA	VSNSE	EEKCL	RWQNE	MRKVG
Equine		APRKS	VRWCT	ISPAE	AAKCA	K FQRN	MKK

FIGURE 3. Comparison of N-terminal sequences of LF molecule from different mammalian species.

The three-dimensional structure of diferric bLF and factors that influence its iron binding were reported (Moore et al., 1997). The final model comprised 5310 protein atoms (residues 5 to 689), 124 carbohydrate atoms (from ten monosaccharide units, in three glycan chains), 2 Fe^{3+}, 2 CO_3^{2-} and 50 water molecules. The folding of bLF molecule was similar to that of hLF, but bovine species differed in the extent of closure of the two domains of each lobe, and in the relative orientations of the two lobes. Differences in domain closure were attributed to amino acid changes in the interface, and differences in lobe orientations to slightly altered packing of two hydrophobic patches between the lobes. Changed inter-domain interactions were implied to the lesser iron affinity of bLF, compared with hLF, and two lysine residues behind the N-lobe iron site of bLF offer new insights into the 'dilysine trigger' mechanism proposed for iron release by TFs. The bLF structure was also notable for several well-defined oligosaccharide units that demonstrate the structural factors that stabilize carbohydrate structure. One glycan chain, attached to *Asn-545*, appears to contribute to inter-domain interactions and possibly modulate iron release from the C-lobe.

C. Heterogeneity

Among mammalian LFs, the human protein has been widely characterized. The amino acid and cDNA sequence data indicate that several animal LFs share extensive regions of primary sequence homology. Specifically, N-terminal sequences for porcine LF (Hutchens et al., 1989) indicate homology between LFs from human (Metz-Boutigue et al., 1984), bovine (Wang et al., 1984), equine (Jolles et al., 1984), monkey (Davidson & Lönnerdal, 1986) and murine (Pentecost & Teng, 1987) origins (*FIGURE 3*). X-ray diffraction studies also have demonstrated certain degrees of structural homology between hLF and bLF molecules (Norris et al., 1986). Peptide mapping also suggested structural homology between porcine and human LFs (Kokriakov et al., 1988).

LFs isolated from various sites of the human body demonstrate antigenic similarity. Considering the total amino acid sequence (Metz-Boutigue et al., 1984) and the polydispersity of the glycan structures (Spik et al., 1982), hLF has an estimated molecular mass of 82,400 $\pm$ 400. However, hLF in a number of human body fluids was found to possess different electrophoretic mobility due to its interaction with acidic macromolecules (Hekman, 1971). Several reports have also suggested polymerization of LF to a variable degree in biological fluids. Different forms of LF seem to appear at distinct stages of certain infections. Tabak et al. (1978) have detected LF polymers in the saliva of a patient with acute parotitis, however, apparent dimers and monomers were recovered when the inflammation gradually subsided. Similarly, bLF trimers appear in milk during acute stages of bovine mastitis, while dimers and eventually monomers emerge as predominant forms during the healing process (Harmon et al., 1976). The LF aggregation phenomenon in calcium containing fluids seems to inactivate certain biological activities of the molecule, such as the feedback control of granulopoiesis (Bennett et al., 1981).

Three isoforms of hLF with identical molecular mass, pI, partial proteolytic peptide patterns, and N-terminal amino acid sequence, but with distinct RNAse activity, were reported. The LF-alpha form binds iron; and, the other two, LF-beta and LF-gamma forms, express potent RNAse activity but lack the iron binding capacity (Furmanski et al., 1989). Two apparent forms of LF were also identified in bovine colostrum and the molecular heterogeneity seems due to a varying degree of protein glycosylation (Tsuji et al., 1989).

V. ANTIMICROBIAL EFFECTS

Structural characteristics and spatial orientation of the molecule are critical factors in the functionality of an antimicrobial compound. Occurrence in various milieu strongly emphasizes the significance of the structure-function relationship in the multifunctionality of the LF. As an exocrine secretory protein LF is present in different biological fluids of varying viscosity, pH, and ionic strength and co-exists with continuously changing ratios of other physiological substances. Thus, LF may be expected to perform a different antimicrobial function in the tear or saliva compared to its activity against an enteric bacteria at the intestinal mucosa. Moreover, as an acute-phase reactant LF also exists as a regulatory molecule in the cellular pool such as the neutrophils, and contributes to antigen processing in the phagosomes. Considering the diversity of LF's role in innate defense, a broad-spectrum of antimicrobial activity is expected. Accordingly, various modes of antimicrobial effects have been reported for LF.

A. Stasis effect

The iron-chelating capacity of LF in the metal-binding pockets in co-ordination with the bicarbonate anion has been suggested in the nutritional deprivation and a consequent inhibition of microbial growth in the stasis effect. This hypothesis was supported with various laboratory findings, such as exogenous addition of iron into the milieu could reverse the stasis effect or iron-saturated LF is non-inhibitory. During the early 1960s, Reiter's laboratory suggested the stasis mechanism of antimicrobial action for LF, which was further substantiated by Masson and co-workers. The stasis effect has been verified and validated by various laboratories during the past three decades. There are vast numbers of peer-reviewed publications in the scientific literature and this section will discuss the salient points.

Kirkpatrick et al. (1971) reported the fungistatic effect of apo-LF against *Candida albicans* and suggested a role for LF in the host-defense mechanism in chronic mucocutaneous candidiasis.

Reiter et al. (1975) found that two strains of *E. coli* were inhibited by colostral whey after dialysis or dilution in Kolmer saline and addition of precolostral calf serum or LF. Undiluted dialyzed milk was not inhibitory due to its low LF content but became inhibitory after addition of 1 mg/ml of LF. The lack of inhibition in undiluted whey is due to the high concentration of citrate in colostral whey (and milk) and it is suggested that citrate competes with the iron-binding proteins for iron and makes it available to the bacteria. Addition of bicarbonate, which is required for the binding of iron by TF and LF, could overcome the effect of citrate.

Bishop et al. (1976) tested the bacteriostatic effects of apo-bLF against strains of coliform bacteria associated with bovine mastitis. As low as 0.02 mg of apo-bLF per ml resulted in marked inhibition of growth of all coliforms. The stasis effect was lost if saturated LF or iron plus apo-LF was added to the synthetic medium. The inhibition of growth increased as the concentration of apo-LF increased from 0.02 to 0.2 mg/ml for *Klebsiella pneumoniae* and 2 mg/ml for *Aerobacter aerogenes*, and *E. coli*. As the concentration of apo-LF was increased above 0.2 or 2 mg/ml, there was less inhibition of growth except for *E. coli*. These results are compatible with the hypothesis that coliform bacteria respond to low-iron environments by production of iron-sequestering agents that compete effectively with apo-LF for free iron. Addition of apo-LF plus citrate resulted in

loss of growth inhibition. The molar ratio (citrate to apo-LF) was found to be more important than the absolute concentration of either component. A ratio of 75 resulted in 50% growth inhibition, whereas ratios of 300 and greater resulted in less than 10% growth inhibition. These results suggest that the ratio of citrate to LF would be important in evaluating LF as a nonspecific protective factor of bovine mammary secretions.

An *in vitro* microassay was developed to evaluate bacteriostatic properties of apo-bLF (Nonnecke & Smith, 1984a). The growth of coliform, staphylococcal, and streptococcal bacterial strains in a defined synthetic medium was inhibited by apo-bLF (0.5 to 30.0 mg/ml). Addition of holo-LF to the synthetic medium did not inhibit growth of test strains. Inhibition by apo-LF was greater for coliform than Gram-positive strains for all concentrations of apo-LF evaluated. No concentration of apo-LF proved bactericidal for either coliform or Gram-positive strains. Inhibition of two coliform strains by apo-LF (10 mg/ml) was abolished by addition of ferric iron to the assay system, indicating an iron-dependent nature of apo-LF induced inhibition of bacteria. Bicarbonate supplementation of the growth system containing apo-LF (1 mg/ml) increased inhibition of three coliform strains by apo-LF. Addition of increasing concentrations of citrate (2.0 mg/ml) to an assay system containing apo-LF (5 mg/ml) resulted in a concomitant reduction of growth inhibition of three coliform strains. These data confirmed a potential relationship between the molar ratio of citrate to LF of the lacteal secretion and its capacity to inhibit coliform strains associated with mastitis.

Mammary secretions were collected during physiologic transitions of the udder and were used in an *in vitro* microbiological assay to determine bacteriostasis of mastitis pathogens (Breau & Oliver, 1986). As mammary involution progressed, *in vitro* stasis of *Klebsiella pneumoniae, E. coli,* and *Streptococcus uberis* increased. Mammary secretions from concanavalin A (conA)- and phytohemagglutinin (PHA)-treated glands had significantly increased bacteriostasis. Secretions contained significantly increased concentrations of LF and a decreased citrate:LF molar ratio earlier in the dry period than did control mammary secretions. Greatest bacteriostasis was observed in mammary secretions obtained 7 days before parturition. However, differences in secretion composition or bacteriostasis were not found between conA- or PHA-treated and control udder halves during the prepartum period. Bacterial growth inhibition by mammary secretion decreased markedly during early lactation. A highly significant positive correlation was found between bacteriostasis and concentrations of LF, serum albumin, and IgG. A highly significant negative correlation was also reported in the citrate:LF molar ratio during early involution and the peripartum period.

The bacteriostasis effect bLF, TF and immunoglobulins against *E. coli* strain B117, acting alone or in combination, was investigated *in vitro* (Rainard, 1986a). Both LF and TF elicited a strong bacteriostasis without requirement for antibodies. After a short period of growth, the multiplication of bacteria was almost completely prevented by the iron-binding proteins. A significant but moderate additional stasis was achieved when IgG or IgM was added to TF, while addition of Ig to LF revealed no significant cooperative effect. All of 11 strains of *E. coli* isolated from bovine mastitis were sensitive to LF in the absence of Ig. It therefore appeared that antibodies were not required for LF to exert a potent bacteriostatic effect on mastitis isolates of *E. coli.* Rainard (1986b) also examined the bacteriostatic activity of bLF against mastitis pathogens using an *in vitro* microassay. The most susceptible species was *E. coli*; all of the 35 isolates tested were susceptible to

bacteriostasis by apo-LF (0.1 mg/ml), although a few strains showed a lower degree of inhibition. Heterogeneity among strains was more pronounced among 10 isolates of *Staphylococcus aureus*, four of which were apparently unaffected by apo-LF (1 mg/ml). Under the same conditions, *Streptococcus agalactiae* (six isolates) and *Strep. uberis* (five isolates) resisted the bacteriostatic action of apo-LF.

The growth of *Streptococcus mutans* 6715-13 in a rich medium (Todd Hewitt broth) was drastically reduced by addition of apo-LF; this effect was bacteriostatic and reversible by saturation of LF with iron (Visca et al., 1989).

Dionysius et al. (1993) examined the *in vitro* antibacterial effects of various forms of LF on enterotoxigenic strains of *E. coli* using a microassay for bacterial growth. Native and apo-LF exhibited variable activity against 19 strains, whereas holo-LF had no effect. At a concentration of .2 mg/ml of apo-LF, strains could be distinguished as either sensitive or resistant to inhibition. Zinc-saturated LF was as bacteriostatic as apo-LF when sensitive and resistant strains were tested over the concentration range .04 to 1.0 mg/ml of LF. A bactericidal effect was observed for native, apo-, and Zn-saturated LF against some sensitive strains. The antibacterial activity of apo-LF depended on bacterial inoculum size and was not enhanced by the addition of lysozyme. Addition of holo-LF or cytochrome c diminished the antibacterial effect of apo-LF, whereas addition of BSA had no effect. Resistance to inhibition by LF was not related to the production of bacterial siderophores.

Paulsson et al. (1993) tested the effect of pasteurization- and UHT-treatments on the LF interaction with bacteria. The ability of native and iron-saturated LF to bind various bacterial species was unaffected by pasteurization. However, UHT treatment decreased this interaction capacity. Native LF, both unheated and pasteurized, showed similar bacteriostatic properties and moderately inhibited *E. coli*. However, this inhibitory capacity was lost after UHT treatment. Iron-saturated LF did not inhibit bacterial growth; neither pasteurization nor UHT could change this property. Thus, UHT seems to affect structural as well as certain biological properties (including bacteriostasis) of both native and iron-saturated bLF, and pasteurization seems to be a treatment of choice for products containing this protein.

The effect of LF on bacterial growth was tested by measuring conductance changes in the cultivation media by using a Malthus-AT system and was compared with the magnitude of ^{125}I-labeled LF binding in 15 clinical isolates of *E. coli* (Naidu et al., 1993). The binding property was inversely related to the change in bacterial metabolic rate and was directly related to the degree of bacteriostasis (*FIGURE 4*). The magnitude of LF-bacterium interaction showed no correlation with the MIC of LF. In certain strains, LF at supraoptimal levels reduced the bacteriostatic effect. Thus, the LF concentration in the growth media was critical for the antibacterial effect. The cell envelopes of *Salmonella typhimurium* 395MS with smooth lipopolysaccharide (LPS) and its five isogenic rough mutants revealed 38-kDa porin proteins as peroxidase-labeled-LF-reactive components in sodium dodecyl suLFate-polyacrylamide gel electrophoresis and Western blot (ligand blot) analysis. However, in the whole cell-binding assay, parent strain 395MS demonstrated a very low interaction with ^{125}I-LF. On the other hand, LF interaction gradually increased in correspondence with the decrease in LPS polysaccharide moiety in the isogenic rough mutants. Conductance measurement studies revealed that the low-level-LF-binding (low-LF-binding) strain 395MS with smooth LPS was relatively insusceptible to LF, while the high-LF-binding mutant Rd was more susceptible to LF (*FIGURE 5*). These

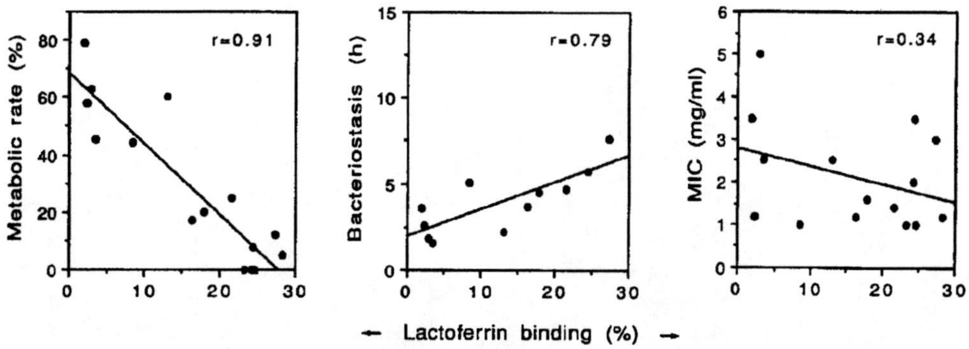

FIGURE 4. LF binding to *E.coli* and its relation to antimicrobial outcome. Correlation between parameters was made by linear regression analysis. The metabolic rate in the presence of LF (1 mg/ml) is expressed as a relative percentage, considering the change in the conductance rate of the control (without LF) as 100%. Bacteriostasis was estimated as the difference between bacteria growth in media with LF and the control. For MIC determinations, bacteria were grown in the presence of 15 different concentrations of LF within a range of 0.01 to 10 mg/ml. The lowest amount of LF that caused complete inhibition of bacterial growth at the time point when the metabolism of the control reached stationary phase was considered as MIC [reproduced from Naidu et al. (1993) with permission from the American Society for Microbiology].

data suggested a correlation between LF binding to porins and the LF-mediated antimicrobial effect. The polysaccharide moiety of LPS shielded porins from the LF interaction and concomitantly decreased the antibacterial effect.

Dial et al. (1998) have examined the *in vitro* and *in vivo* antimicrobial efficacy of bLF against *Helicobacter* species. LF was bacteriostatic to *H. pylori* when cultured at concentrations above or 0.5 mg/ml. Growth of *H. pylori* was not inhibited by another milk constituent, lysozyme, or by bLFcin, but growth was inhibited by the iron chelator deferoxamine mesylate. LF inhibition of growth could be reversed by addition of excess iron to the medium. LF in retail dairy milk was found to be more stable intra-gastrically than unbuffered, purified LF. Treatment of *H. felis*-infected mice with LF partially reversed mucosal disease manifestations. The authors concluded that bLF has a significant antimicrobial activity against *Helicobacter* species *in vitro* and *in vivo*.

The growth of *Bacillus cereus* was markedly inhibited by the addition of LF and was recovered by the addition of $FeCl_3$ (Sato et al., 1999). The bacteriostasis was also reversed by the addition of erythrocytes and hemoglobin. *B. cereus* could use heme or heme-protein complex (hemoglobin-haptoglobin and hematin-albumin complexes) as iron sources in iron deficient conditions. Thus, *B. cereus* seems to use such heme or heme-protein complexes to prevent the bacteriostasis of LF *in vivo*.

B. Cidal effect

In 1977, Arnold and co-workers reported a bactericidal effect for native LF molecule, which apparently was distinct from the stasis mechanism. These experiments were performed with microbial cells suspended in deionized water or buffer solutions at acid pH and the reported mechanism is controversial. Various laboratories have failed to demonstrate a similar bactericidal effect (Rainard, 1987; Gutteberg et al., 1990). Lassiter (1990), from Arnold's group, later published a doctoral thesis which indicated that cont-

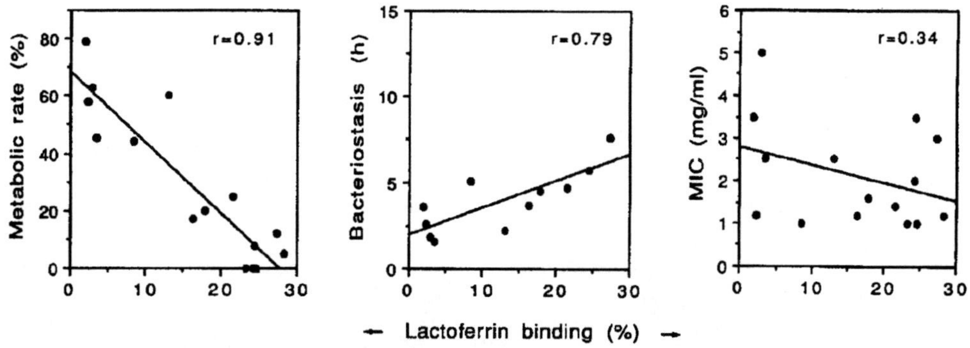

FIGURE 4. LF binding to *E.coli* and its relation to antimicrobial outcome. Correlation between parameters was made by linear regression analysis. The metabolic rate in the presence of LF (1 mg/ml) is expressed as a relative percentage, considering the change in the conductance rate of the control (without LF) as 100%. Bacteriostasis was estimated as the difference between bacteria growth in media with LF and the control. For MIC determinations, bacteria were grown in the presence of 15 different concentrations of LF within a range of 0.01 to 10 mg/ml. The lowest amount of LF that caused complete inhibition of bacterial growth at the time point when the metabolism of the control reached stationary phase was considered as MIC [reproduced from Naidu et al. (1993) with permission from the American Society for Microbiology].

data suggested a correlation between LF binding to porins and the LF-mediated antimicrobial effect. The polysaccharide moiety of LPS shielded porins from the LF interaction and concomitantly decreased the antibacterial effect.

Dial et al. (1998) have examined the *in vitro* and *in vivo* antimicrobial efficacy of bLF against *Helicobacter* species. LF was bacteriostatic to *H. pylori* when cultured at concentrations above or 0.5 mg/ml. Growth of *H. pylori* was not inhibited by another milk constituent, lysozyme, or by bLFcin, but growth was inhibited by the iron chelator deferoxamine mesylate. LF inhibition of growth could be reversed by addition of excess iron to the medium. LF in retail dairy milk was found to be more stable intra-gastrically than unbuffered, purified LF. Treatment of *H. felis*-infected mice with LF partially reversed mucosal disease manifestations. The authors concluded that bLF has a significant antimicrobial activity against *Helicobacter* species *in vitro* and *in vivo*.

The growth of *Bacillus cereus* was markedly inhibited by the addition of LF and was recovered by the addition of $FeCl_3$ (Sato et al., 1999). The bacteriostasis was also reversed by the addition of erythrocytes and hemoglobin. *B. cereus* could use heme or heme-protein complex (hemoglobin-haptoglobin and hematin-albumin complexes) as iron sources in iron deficient conditions. Thus, *B. cereus* seems to use such heme or heme-protein complexes to prevent the bacteriostasis of LF *in vivo*.

B. Cidal effect

In 1977, Arnold and co-workers reported a bactericidal effect for native LF molecule, which apparently was distinct from the stasis mechanism. These experiments were performed with microbial cells suspended in deionized water or buffer solutions at acid pH and the reported mechanism is controversial. Various laboratories have failed to demonstrate a similar bactericidal effect (Rainard, 1987; Gutteberg et al., 1990). Lassiter (1990), from Arnold's group, later published a doctoral thesis which indicated that cont-

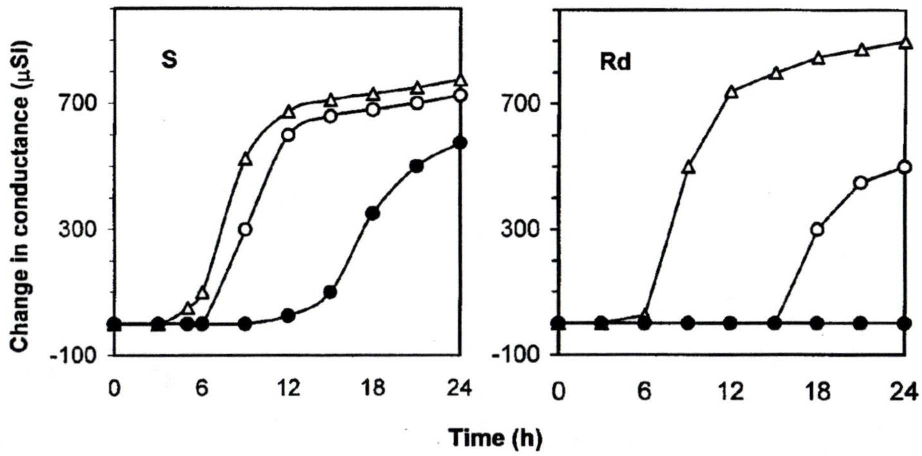

FIGURE 5. Growth of *S. typhimurium* 395MS(S) (S; parent with smooth LPS) and its isogenic mutant Rd (with rough LPS) in SPYE broth. Control (△), with 1 mg/ml bLF (○) and 5 mg/ml bLF (●). [Redrawn from Naidu, et al., 1993].

amination of EDTA during dialysis of LF could account for the cidal effect against *E.coli*. Furthermore, the cidal effects of LF against oral streptococci seem due to the acid pH of the test system. Degradation products in an LF preparation such as the cationic peptides could elicit membrane damage and kill microorganisms. In an antimicrobial milieu such as in the phagosome, LF could possibly elicit a cidal effect synergistic with oxidative events. However, clear evidence for a direct cidal effect with native (intact) LF molecule *in vitro* is still lacking. This section has reviewed the cidal effect in a chronological perspective with no endorsement for the mechanisms hypothesized in the literature.

LFs seem to elicit cidal effects against a variety of microorganisms including Gram-positive and Gram-negative bacteria, rods and cocci, facultative anaerobes, and aero-tolerant anaerobes. Similar morphological and physiological types are represented among the LF-resistant bacteria (Arnold et al., 1980). *S. mutans* was more resistant to LF when grown on a sucrose-containing medium than when it was grown on brain heart infusion broth without added sucrose. When an LF-sensitive, avirulent strain of *Streptococcus pneumoniae* was passed through mice, the resultant virulent culture demonstrated resistance to LF. Since organisms of the same species and even of the same strain such as *S. pneumoniae*, can differ in susceptibility to LF, it appears that accessibility to the LF target site may account for variations in susceptibility.

Influence of several physical conditions and the metabolic state of *Streptococcus mutans* on LF susceptibility were reported (Arnold et al., 1981). After exposure to LF, a 15-min lag period occurred before the initiation of killing, indicating that a two-step process is involved in LF killing. Cultures harvested during the early exponential phase were sensitive to LF, whereas cultures harvested in the early stationary phase were markedly resistant. The rate of killing was dependent on temperature; there was no loss of viability at 2°C. Killing occurred at pH 5.0 to 6.0 in water and 20 mM glycine, but not at any pH in 50 mM sodium phosphate or N-2 hydroxyethylpiperazine-N'-2-ethanesulfonic acid (HEPES) buffer. Addition of exogenous ferrous or ferric ions did not reverse or prevent LF killing, nor did 1 mM magnesium chloride.

Bactericidal effect of LF against *Legionella pneumophila* was reported (Bortner et al., 1986). Purified apo-hLF elicited cidal activity against *L. pneumophila* (serogroup 1), with a 4-log decrease in viability within 2 h at 37 °C. Guinea pig passage of this strain did not affect its sensitivity to LF. Addition of magnesium blocked the bactericidal activity. In addition, human milk was also cidal for *L. pneumophila.* Salts including $CaCl_2$, $Mg(NO_3)_2$, and $MgCl_2$, but not NaCl, blocked killing. Activity was pH dependent with the greatest activity at 5.0. Sensitivity of the organism was markedly affected by the growth conditions. Log-phase 12 h, broth-grown cells were most sensitive, with older cultures appearing more resistant. Plate-grown cells were completely resistant. LF binding, as detected by immunofluorescence microscopy, was temperature dependent (no binding was observed at 4°C), but was independent of killing (Bortner et al., 1989).

Actinobacillus actinomycetemcomitans is a fastidious, facultative Gram-negative rod associated with endocarditis, certain forms of periodontal disease, and other focal infections. Human LF is bactericidal for this pathogen (Kalmar & Arnold, 1988). This cidal activity required an unsaturated (iron- and anion-free) molecule that produced a 2-log reduction in viability within 120 min at 37°C at a concentration of 1.9 μM. Magnesium enhanced LF killing, while other cations, such as potassium and calcium, had no effect.

It was reported that selective anions were capable of inhibiting the expression of bactericidal activity by LF on *S. mutans* 10449 (Lassiter et al., 1987). The ability to block LF expression was directly related to the capacity of the anion to serve as a coordinate ion in iron-binding by the LF molecules. The authors hypothesized the presence of an anionic LF target site on the bacterial surface. Treatment of *S. mutans* with LF under anaerobic conditions abrogated the bactericidal effect. LF killing could be enhanced with thiocyanate and inhibited by catalase and lactoperoxidase; however, bovine serum albumin was equally effective as an inhibitor.

Antimicrobial effects of LF and human milk on *Yersinia pseudotuberculosis* was reported (Salamah & al-Obaidi, 1995a). Bacterial growth *in vitro* was inhibited by apo- but not holo-LF or human milk. Iron-free human milk and to a lesser extent normal human milk were bactericidal for *Y. pseudotuberculosis* cells that were suspended in deionized water. The *in vivo* studies also showed that iron-saturated LF enhanced growth, whereas the viable count was reduced by iron-free LF and EDTA. Nine envelope proteins were decreased or disappeared upon growth in iron-deficient medium, whereas one new high molecular weight protein appeared under the same conditions. The effect of pH, temperature, concentration of magnesium and calcium on the bactericidal activity of LF against *Yersinia pseudotuberculosis* was investigated (Salamah & al-Obaidi, 1995b). The bactericidal activity of LF was higher at acid pH, whereas the bactericidal activity of TF was higher at alkaline pH. Both were not efficient at 4°, 15°, and 25°C, but were efficient at 37°C. LF, but not TF, was very efficient at 42°C. The activity of both proteins were time and concentration dependent. Calcium did not effect their activity up to 60 mM, whereas magnesium reduced the activity of LF only.

C. Adhesion-blockade effect

E. coli is one of the major etiological agents of gastrointestinal illnesses in humans and animals. Bacterial adherence to intestinal epithelia is an important step in the pathogenesis of this disease. In the colonization process, bacterial adhesins such as fimbriae may recognize various mammalian subepithelial matrix components as receptors.

Substances that interfere in this host-pathogen interaction could be of therapeutic and prophylactic value, and nonimmunoglobulin fractions of milk, ileal mucus and mucin are among such potential inhibitors (Holmgren et al., 1981; Miedzobrodzki et al., 1989; Olusanya & Naidu, 1991; Cravito et al., 1991).

1. Adhesion-blockade of enteric pathogens. Several carbohydrates, such as 0.1% fucose or 0.5% glucose, as well as LPS (10 μg/ml) isolated from *Shigella flexneri* strongly inhibit the adherence of shigellae to guinea pig colonic cells. Fucose-containing peptides from hLF also inhibit the adhesion of *S. flexneri* to colonic epithelial cells (Izhar et al., 1987)

The non-immunoglobulin component of human milk responsible for the inhibition of *E. coli* cell adhesion (hemagglutination) mediated by colonization factor antigen I (CFA-I) were identified by chromatographic fractionation of human whey proteins (Giugliano et al., 1995). Free secretory component (fSC) and LF were isolated and both compounds inhibited the hemagglutination by *E. coli* CFA1+. The lowest concentrations of fSC and LF able to inhibit the hemagglutination by *E. coli* strain TR50/3 CFA1+ were 0.06 mg/ml and 0.1 mg/ml, respectively. Commercial preparations of LF from human milk and TF from human serum also inhibited the hemagglutination, with MIC values of 0.03 mg/ml and 0.4 mg/ml, respectively.

Bovine LF mediated inhibition of hemagglutination activity of type 1 fimbriated *E. coli* has also been reported (Teraguchi et al., 1996). The agglutination reaction was specifically inhibited by glycopeptides derived from bLF or α-methyl-D-mannoside. These observations indicate that the glycans of bLF could serve as receptors for type 1 fimbrial lectin of *E. coli*.

The ability of LF to inhibit *in vivo* colonization of *E. coli* has been examined (Naidu et al., unpublished). Infection with *E. coli* strain F18 was established in streptomycin-treated mice by gastric intubation and bacterial excretion was estimated as colony forming units per gram (CFU/g) feces. The excretion of strain F18 in feces reached a steady-state (10^8 CFU/g) within 7 days, independent of challenge (dose: 8 x 10^8 or 10^3 CFU). Oral administration of bLF (20 mg/ml in 20% sucrose solution) caused a 1- to >3-log reduction in CFU/g feces with high and low dosages of strain F18. The bacterial multiplication *in vivo* was markedly affected during the early 24 hours of infection, reflecting >3-log lower number of bacteria in the feces (2 x 10^3 CFU/g) than the control group. Oral administration of LF prior to infection reduced fecal excretion of *E. coli* from mouse intestine. *In vitro* effects of bLF on the molecular interactions of *E. coli* with subepithelial matrix proteins were examined. Bovine LF inhibited the binding of ^{125}I-labeled fibronectin, fibrinogen, collagen type-I, collagen type-IV and laminin to bacteria. This inhibitory effect was bLF dose-dependent, and was independent of coexistence (competitive) or preexistence (non-competitive) of bLF with the tissue matrix proteins. In displacement studies with bacteria-matrix protein complexes, bLF dissociated only collagen type-I and laminin interactions. Electron microscopy revealed the loss of type-1, CFA-I and CFA-II fimbria of *E. coli* grown in broth containing 10 μM LF. The inhibitory affect of LF on fimbrial expression was further confirmed by hemagglutination and yeast cell agglutination. The presence of 10 μM LF in the growth media, however, did not affect the P-fimbriation in *E.coli*. These data suggest a strong influence of LF on adhesion-colonization properties of *E.coli*.

2. Adhesion-blockade of oral pathogens. The influence of LF, salivary proteins (SP) and BSA on the attachment of *Streptococcus mutans* to hydroxyapatite (HA) was reported (Visca et al., 1989). Sorption of LF, SP, and BSA to HA was dependent on the protein concentration and reached the end-point at about 80 mg of proteins per gram of HA. Similarly, the number of streptococci adsorbed to HA was correlated to the amount of cells available up to at least 10^7 cells per mg of HA. The adsorption of LF, SP and BSA on HA reduced the number of attaching *S. mutans* cells. In particular, SP reduced the adsorption of *S. mutans* by 30%, whereas pre-coating of HA with apo- or iron-saturated LF resulted in a three orders of magnitude reduction of *S. mutans* adsorption to HA. The potent adherence-inhibiting effect of apo-LF together with its antibacterial activity against *S. mutans* suggests an important biological significance of these phenomena in the oral cavity.

Whole cells of *P. intermedia* demonstrate a high degree of binding to fibronectin, collagen type I and type IV and laminin, whereas a moderate interaction was detected with fibrinogen. The ability of bLF to affect the interactions of the above proteins with *P. intermedia* was examined (Alugupalli et al., 1994). In the presence of unlabeled bLF, a dose-dependent inhibition of binding was observed with all five proteins tested. Unlabeled bLF also dissociated the bacterial complexes with these proteins. The complexes with laminin or collagen type I were more effectively dissociated than fibronectin or fibrinogen, whereas the interaction with collagen type IV was affected to a lesser extent. A strain-dependent variation in the effect of bLF was observed.

The ability of hLF and bLF to inhibit adhesion of *A. actinomycetemcomitans* and *P. intermedia* to monolayers of fibroblasts, HEp-2, KB and HeLa cells was reported (Alugupalli & Kalfas, 1995). The inhibitory effect was dose-dependent in the concentration range 0.5-2500 μg/ml and not related to the bacterial growth phase. In the presence of LF, decreased association of bacteria with the cell monolayers was also found by microscopic examination of the preparations. These data suggested a possibility that LF could prevent the establishment of bacteria in periodontal tissues through adhesion-counteracting mechanisms in addition to its bacteriostatic and bactericidal properties.

LF also binds to fibroblast monolayers and matrigel, a reconstituted basement membrane, through ionic interactions. The adhesion of *A. actinomycetemcomitans* to these substrata was mainly dependent on the ionic strength of the environment. *P. intermedia* and *P. nigrescens* also adhere to fibroblasts mainly by ionic interactions, while their adhesion to matrigel seems to be mediated by specific mechanisms. Lectin-type interactions were not found involved in the binding of these bacteria to the substrata. Treatment of either *A. actinomycetemcomitans* or fibroblasts with LF decreased the adhesion in a dose-dependent manner, while LF treatment of matrigel alone had no adhesion-counteracting effect. Adhesion of *P. intermedia* and *P. nigrescens* to matrigel was not significantly affected by the ionic strength, but the presence of LF inhibited the adhesion. LF bound to matrigel, *P. intermedia* and *P. nigrescens* was rapidly released, while LF bound to *A. actinomycetemcomitans* and fibroblasts was retained. These findings indicate that LF-dependent adhesion-inhibition of *A. actinomycetemcomitans, P. intermedia* and *P. nigrescens* to fibroblasts and matrigel could involve binding of LF to both the bacteria and substrata. The decreased adhesion may be due to blocking of both specific adhesin-ligand as well as non-specific charge-dependent interactions (Alugupalli & Kalfas, 1997).

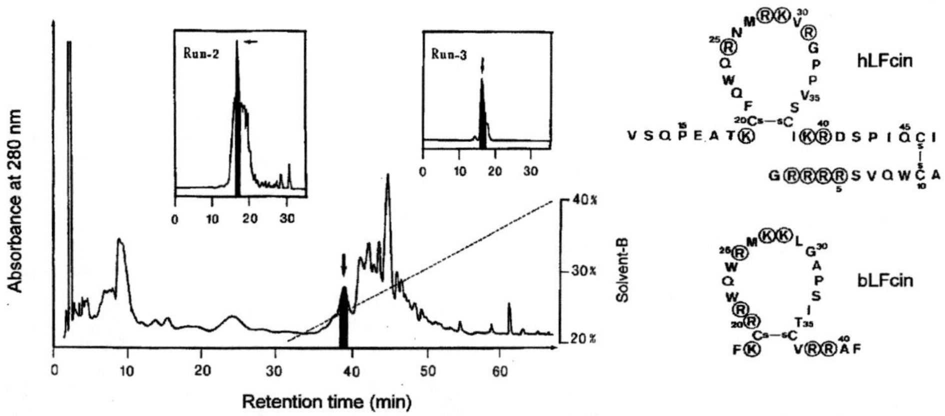

FIGURE 6. LFcin isolation by reverse-phase HPLC. Bovine LF was hydrolyzed with porcine gastric pepsin at pH 4.0 and the hydrolysate was fractionated on a Pep-S column. Shaded peaks are fractions with antimicrobial activity against *E.coli* H10407 in a microplate assay [Naidu & Erdei, unpublished data]. The amino acid sequence and the primary structures of bLFcin and hLFcin peptides are shown with basic residues encircled and sequence positions numbered [adapted from Bellamy et al., 1992].

D. Cationic effect

LF was found to contain an antimicrobial sequence near its N-terminus, which appears to function by a mechanism distinct from iron chelation. The identified domain contains a high proportion of basic residues, like various other antimicrobial peptides known to target microbial membranes and it appears to be located on the surface of the folded protein allowing its interaction with surface components of microbial cells (Tomita et al., 1994).

Hydrolysates prepared by cleavage of bLF with porcine pepsin, cod pepsin, or acid protease from *Penicillium duponti* showed strong activity against *E. coli* O111, whereas hydrolysates produced by trypsin, papain, or other neutral proteases were much less active (Tomita et al., 1991). Low molecular weight peptides generated by porcine pepsin cleavage of bLF showed broad-spectrum antibacterial activity, inhibiting the growth of a number of Gram-negative and Gram-positive species, including strains that were resistant to native LF. The antibacterial potency of the hydrolysate was at least eight-fold greater than that of undigested LF with all strains tested. The active peptides retained their activity in the presence of added iron, unlike native LF. The effect of the hydrolysate was bactericidal as indicated by a rapid loss of viability of *E. coli* O111.

A single active peptide representing antimicrobial domain was isolated following gastric pepsin cleavage of hLF, and bLF, and sequenced by automated Edman degradation. The antimicrobial sequence was found to consist mainly of a loop of 18 amino acid residues formed by a disulfide bond between cysteine residues 20 and 37 of hLF, or 19 and 36 of bLF (Bellamy et al., 1992). Synthetic analogs of this region similarly exhibited potent antibacterial properties. The active peptide of bLF was more potent than that of hLF having effectiveness against various Gram-negative and Gram-positive bacteria at concentrations between 0.3 μM and 3.0 μM, depending on the target strain. Effect of the isolated domain was lethal causing a rapid loss of colony-forming capability (*FIGURE 6*).

Human LF contains a 46 residue sequence named lactoferricin H (hLFcin) responsible for its cationic antimicrobial properties. Synthetic peptides HLT1, corresponding to the loop region of hLFcin (*FQWQR-NMRKVRGPPVS*) and HLT2, corresponding to its charged portion (*FQWQRNMRKVR*), exerted significant antibacterial effects against *E. coli* serotype O111 strains NCTC 8007 and ML35 (Odell et al., 1996). The corresponding sequences in native hLF were shown to adopt a charged helix and hydrophobic tail within the N-lobe remote from the iron binding site. Sequence similarities between LFcin and dermaseptin and magainins suggest that LFcin may act as an amphipathic α-helix.

The basic amino acid-rich region of bovine lactoferricin (bLFcin), *RRWQWRMKKLG* has many basic and hydrophobic amino acid residues. Using chemically synthesized bLFcin and its substituted peptides, the antimicrobial activities of the peptides were tested by determining the minimal inhibitory concentration (MIC) of *E. coli* and *Bacillus subtilis* and the disruption of the outer cell membrane of *E. coli,* and the peptide's toxicities were assayed by hemolysis (Kang et al., 1996). The short peptide (B3) composed of only 11 residues had similar antimicrobial activities while losing most of the hemolytic activities as compared with the 25 residue-long ones (B1 and B2). The short peptides (B3, B5 and B7) with double arginines at the N-termini had more potent antimicrobial activity than those (B4 and B6) with lysine. However, no antimicrobial and hemolytic activities were found in B8, in which all basic amino acids were substituted with glutamic acid, and in B9, in which all hydrophobic amino acids were substituted with alanine. The circular dichroism (CD) spectra of the short peptides in 30 mM SDS were correlated with their antimicrobial activities. These results suggested that the 11-residue peptide of bLFcin is involved in the interaction with bacterial phospholipid membranes and may play an important role in antimicrobial activity with little or no hemolytic activity.

To study the immunochemical and structural properties of bLFcin derived from N-lobe of bLF, monoclonal antibody (mAb) was prepared and the amino acid sequence concerned with binding to mAb identified (Shimazaki et al., 1996). Mice injected with bLFcin showed no production of antibody specific to this peptide, whereas those with bLFcin-KLH conjugate produced anti-bLFcin antibodies. None of the mAb reacted with bLF C-lobe, hLF or hLFcin. By the reactivity of the mAb against the peptides synthesized on cellulose membranes using spots and against chemically modified derivatives of bLFcin, the antigenic determinant was identified to be the sequence '*QWR*'.

Furthermore, three peptides with antibacterial activity toward enterotoxigenic *E. coli* have been purified from a pepsin digest of bLF (Dionysius & Milne, 1997). All peptides were cationic and originated from the N-terminus of the molecule in a region where a bactericidal peptide, bLFcin, had been previously identified. The most potent peptide, peptide I, was almost identical to bLFcin; the sequence corresponded to residues 17 to 42, and the molecular mass was 3195 as determined by mass spectrometry. A second, less active peptide, peptide II, consisted of two sequences, residues 1 to 16 and 43 to 48 (molecular mass of 2673), linked by a single disulfide bond. The third peptide, peptide III, also a disulfide-linked hetero-dimer, corresponded to residues 1 to 48 (molecular mass of 5851), cleaved between residues 42 and 43. Peptides I and II displayed antibacterial activity toward a number of pathogenic and food spoilage microorganisms, and peptide I inhibited the growth of *Listeria monocytogenes* at concentrations as low as 2 μM. Bacterial growth curves showed that bactericidal effects of peptides I and II were observable with-

in 30 min of exposure. The results confirmed and extended those of earlier studies suggesting that the bactericidal domain of LF was localized in the N-terminus and did not involve iron-binding sites.

However, the antibacterial studies conducted by Hoek and co-workers (1997) indicated that the activity of LFcin is mainly, but not wholly, due to its N-terminal region. Several peptides sharing high sequence homology with bLFcin were generated from bLF with recombinant chymosin. Two peptides were co-purified, one identical to bLFcin and another differing from this cationic peptide by the inclusion of a C-terminal alanine. Two other peptides were copurified from chymosin-hydrolyzed LF, one differing from bLFcin by the inclusion of C-terminal alanyl-leucine and the other being a heterodimer linked by a disulfide bond. These peptides were isolated in a single step from chymosin-hydrolyzed LF by membrane ion-exchange chromatography and were purified by reverse-phase high-pressure liquid chromatography (HPLC). They were characterized by N-terminal Edman sequencing, mass spectrometry, and antibacterial activity determination. Pure LFcin, prepared from pepsin-hydrolyzed LF, was purified by standard chromatography techniques. This peptide was analyzed against a number of Gram-positive and Gram-negative bacteria before and after reduction of its disulfide bond or cleavage after its single methionine residue and was found to inhibit the growth of all the test bacteria at a concentration of 8 μM or less. Sub-fragments of LFcin were isolated from reduced and cleaved peptide by reverse-phase HPLC. Sub-fragment 1 (residues 1 to 10) was active against most of the test microorganisms at concentrations of 10 to 50 μM. Sub-fragment 2 (residues 11 to 26) was active against only a few microorganisms at concentrations up to 100 μM.

E. Synergistic effect

LF in combination with antibodies has powerful bacteriostatic effects *in vitro* and this phenomenon provides specific protection against many infections. LF appears to be essential for the antimicrobial function of polymorphonuclear leukocytes against *Pseudomonas aeruginosa* (Bullen, 1981).

Ellison et al. (1990) studied the ability of LF and TF to damage the Gram-negative outer membrane. Lipopolysaccharide (LPS) release by the proteins could be blocked by concurrent addition of Ca^{2+} and Mg^{2+}. Addition of Ca^{2+} also blocked the ability of LF to increase the susceptibility of *E. coli* to rifampicin. TF, but not LF, increased susceptibility of Gram-negative bacteria to deoxycholate, with reversal of sensitivity occurring with exposure to Ca^{2+} or Mg^{2+}. In transmission electron microscopy studies polymyxin B caused finger-like membrane projections, but no morphological alterations were seen in cells exposed to EDTA, LF or TF. These data provide further evidence that LF and TF act as membrane-active agents with the effects modulated by Ca^{2+} and Mg^{2+}.

Antimicrobial activities of LF were tested against 15 strains of 10 species of bacteria, and potent activities against *Staphylococcus aureus, E. coli, Klebsiella pneumoniae* and *Proteus spp.* were observed. Concomitant use of LF with antibiotic cefodoxime proxetil resulted in a synergistic activity against *S. aureus, E. coli, K. pneumoniae* and *Pseudomonas aeruginosa*; and an additive activity against *E. coli* strain NIHJ and *Providencia rettgeri*. The minimum inhibitory concentrations (MICs) of antibiotic in the presence of LF was reduced to < 1/64 with an efficacy rate of 53/57 (92.9%) in a patient group with infections (Chimura et al., 1993).

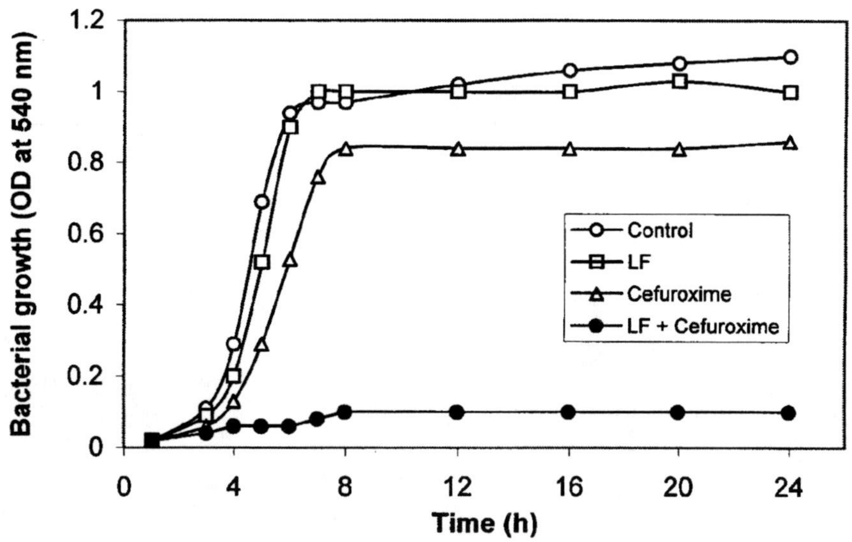

FIGURE 7. LF synergism with antibiotic cefuroxime. Growth of *S. typhimurium* strain ATCC13311 in special peptone yeast extract broth at 37°C was measured as change in optical density at 540 nm. Bacterial growth in media: control (○); 500 μg/ml of bLF (□); 0.25 g/ml cefuroxime (△) and mixture of both agents at the above concentrations (●) [from Naidu & Arnold (1994) with permission from Elsevier Science Inc.].

The antibiotic susceptibility of *Salmonella* sp. in the presence of LF was examined (Naidu & Arnold, 1994). A mixture containing sub-minimum inhibitory concentration levels of LF (MIC/4) and cefuroxime (MIC/2) inhibited the bacterial growth. LF strongly potentiated the action of erythromycin (eight-fold), whereas it increased the activity only by two-fold for ampicillin, ciprofloxacin, chloramphenicol, and rifampicin; similarly, these antibiotics also reduced the MIC of bLF by two-fold in *S. typhimurium*. Such antimicrobial potentiation was not observed with bLF mixtures containing cefalexin, gentamycin, or polymyxin B against strain ATCC13311. BLF and cefuroxime also demonstrated potentiation of varying degrees (2- to 16-fold) with nine other *Salmonella* species. These data established the binding of LF to porins in salmonellae and a potentiation effect of LF with certain antibiotics (*FIGURE 7*).

The effects of LF and its peptides in combination with azole antifungal agents against *Candida albicans* were investigated by a micro-broth-dilution method (Wakabayashi et al., 1996). For pepsin hydrolysate of LF or the LF-derived antimicrobial peptide bLFcin, the concentrations required to inhibit the growth of *Candida* decreased in the presence of relatively low concentrations of clotrimazole (CTZ). The MIC of all azole antifungal agents tested was reduced by 1/4-1/16 in the presence of a sub-MIC level of each of these LF-related substances. Polyene and fluoropyrimidine antifungal agents did not show such a combined effect with these LF-related substances. The anti-*Candida* activity of LF or bLFcin in combination with CTZ was shown synergistic by checkerboard analysis. These results indicate that LF-related substances function cooperatively with azole antifungal agents against *C. albicans*.

Effects of apo-LF and lactoperoxidase system (lactoperoxidase, LP/SCN-/H_2O_2), separately and together, on the viability of *Streptococcus mutans* (serotype c) *in vitro* was tested (Soukka et al., 1991). Streptococci were incubated in buffered KCl (pH 5.5) with

and without the above components at concentrations normally present in human saliva. Both apo-LF and LP-system had a bactericidal effect against *S. mutans* at low pH. Together they showed an additive, but not a synergistic, antibacterial effect against *S. mutans*. Apo-LF enhanced the LP enzyme activity but decreased the yield of the antimicrobial component, hypothiocyanite (HOSCN/OSCN-), when incorporated into the reaction mixtures. This decrease, which was most pronounced at low pH, was due to an LP-independent reaction between apo-LF and HOSCN/OSCN-.

The effect of an antimicrobial protein, calprotectin, in combination with neutrophils on the growth of *C. albicans* was investigated (Okutomi et al., 1998). The growth inhibition of *C. albicans* by murine neutrophils was augmented by the addition of a low concentration of calprotectin prepared from rat peritoneal exudate cells. The concentrations of calprotectin causing 50% inhibition of growth of *C. albicans* in the absence or presence of neutrophils at an effector-to-target (E/T) ratio of 30 and 60 were estimated to be 0.45, 0.34 and 0.28 U/ml, respectively. The anti-*Candida* activity of calprotectin was completely inhibited by 2 μM of zinc ion, while it only partially lowered the activity of the combination of calprotectin and neutrophils. LF has strongly inhibited the growth of *C. albicans* in combination with calprotectin. These results suggest that calprotectin and LF released from neutrophils may cooperate to inhibit the growth of *C. albicans* at a local lesion of the infection where there is an accumulation of neutrophils.

F. Opsonic effect

The ability of hLF to stimulate the phagocytic and cytotoxic properties of macrophages was reported (Lima & Kierszenbaum, 1987). Fe-LF molecule was not required to increase the capacity of mouse peritoneal macrophages to take up *Trypanosoma cruzi* amastigotes, *Listeria monocytogenes*, or latex particles; it was necessary for LF to enhance intracellular killing of the two microorganisms. Thus, apo-LF, which did not increase macrophage cytotoxicity, after restoration of ferric ions prior to its use in treatments or when ferric citrate was added to the culture medium immediately after apo-LF treatment of the macrophages, does increase macrophage cytotoxicity. In that iron ions cannot be internalized as such, the latter observation suggested that apo-LF had taken up iron while membrane bound and then enhanced killing. Immunofluorescence studies revealed that comparable proportion of macrophage-bound apo-LF or LF at either 20 or 100% iron saturation were without appreciable differences in fluorescence intensity. Therefore, reduced binding of apo-LF compared with LF was not a likely explanation for the lack of effect of apo-LF on macrophage killing. LF did not enhance amastigote killing by macrophage in the presence of the iron chelator deferoxamine. Diethylaminetriamine-pentaacetic acid, an iron chelator which is not incorporated into cells, had a similar effect. The iron-binding protein TF did not alter the capacity of macrophage to either take up or kill the amastigote, indicating that the noted LF effects were not shared by all iron-binding proteins. However, prior treatment of macrophages with TF enabled the cells to display a greater parasite killing capacity after apo-LF treatment, suggesting a role for iron in this activity.

Among the known life cycle stages of *Trypanosoma cruzi* only the amastigote form binds LF. This capacity was readily demonstrable by indirect immunofluorescence in amastigotes derived from mice, a mammalian cell culture, or grown in an axenic medium. No LF binding was detectable on trypomastigotes from blood or mammalian cells, insect derived metacyclics or epimastigotes, or on epimastigotes grown in Warren's

medium. Serum levels of LF were increased in mice acutely infected with *T. cruzi*, and amastigotes from the spleens of these animals were found to have the glycoprotein on their surface. The amastigote LF receptor may have biological significance in parasite-host interaction since mononuclear phagocytes also express an LF receptor, and treatment of these cells with LF has been shown to increase their capacities to take up and kill *T. cruzi* amastigotes *in vitro*. The LF receptor is the first marker for *T. cruzi* amastigotes for which a naturally occurring ligand has been described (Lima et al., 1988).

LF bound to *Streptococcus agalactiae* could interfere with the deposition of complement components on the bacterial surface (Rainard, 1993). Pretreatment of streptococci with LF shortened the lag phase preceding the deposition of C3 component on bacteria. The kinetics of C3 deposition was comparable to that obtained by adding antibodies against *S. agalactiae* to agammaglobulinaemic precolostral calf serum (PCS) heated at 56 °C for 3 min to inactivate the alternative pathway. Accelerated C3 deposition did not occur in the absence of calcium ions. Deposition of C4 on bacteria occurred only when either antibodies or LF were added to PCS. These results demonstrate that the interaction of LF with bacteria activated the classical pathway of complement in the absence of antibodies. The binding of purified C1q to bacteria was promoted in a dose-dependent manner by LF, suggesting that recruitment of classical pathway of complement resulted from the interaction of C1q with LF-adsorbed to the bacterial surface. Phagocytosis of bacteria opsonized with heated PCS (at 56°C for 3 min) and LF was comparable to that occurring in the presence of heated PCS and antibodies. These data suggested that LF was able to substitute for antibodies in order to activate the classical pathway of complement and to opsonize unencapsulated *S. agalactiae* efficiently.

VI. ANTIMICROBIAL SPECTRUM

Both LF-susceptible and -resistant organisms encompass a variety of types including Gram-positive and Gram-negative bacteria, rods and cocci, and aerobes and anaerobes; both DNA and RNA viruses; a variety of yeast as well as fungi; and parasites. Susceptibility depends on similarities in cell surface structure or the mode of LF action against individual organisms.

A. Antibacterial activity

The antibacterial properties of milk have been observed for a long time. Most of the relevant literature consists of observations that various pathogenic and saprophytic bacteria are killed or their growth temporarily inhibited by cow's milk. Bacteriostasis was the widely characterized inhibitory mechanism of LF with well-documented data. Over the past three decades, various laboratories have identified LF as a broad-spectrum antimicrobial and reported a variety of inhibitory mechanisms on both Gram-positive and Gram-negative bacteria (*TABLE 3*).

1. Gram-positive bacteria. In 1967, Reiter and Oram reported the antibacterial effects of LF against *Bacillus stearothermophilus* and *B. subtilis*. This study also observed that apo-LF was unable to inactivate bacterial spores but could inhibit their germination. A decade later, Arnold and co-workers (1977) reported cidal activity of LF against *Streptococcus mutans* and other oral streptococci.

The occurrence of LF in saliva has initiated many studies on antimicrobial activity against oral streptococci and control of caries. Apo-LF could cause a potent *in vitro* growth-inhibition of *Streptococcus mutans* and this effect could be reversed by iron (Visca et al., 1989). Furthermore, LF seems to reduce the adsorption of *S.mutans* cells to hydroxyapatite. This adherence-inhibiting effect of apo-LF together with bacteriostasis activity towards *S. mutans* suggests a possible patho-biological significance of caries control in the oral cavity *in vivo*. However, apo-bLF seems to elicit a low degree antimicrobial effect on mastitis-associated streptococci in bovine mammary secretions (Todhunter et al., 1985).

Naidu and co-workers (1990; 1991) have identified specific LF-binding proteins in *Staphylococcus aureus* isolated from human and animal infections as well as among various species of coagulase-negative staphylococci causing bovine mastitis. Apo-bLF at concentrations of 0.1%-0.4% could convert compact colonies of *Staphylococcus haemolyticus* transient to diffused in soft agar (Godo et al., 1997). This surface-active property of LF has prevented autoaggregation of cocci in compact ball-like colonies by hydrophobic interaction or trypsin-sensitive proteins. *In vivo* anti-staphylococcal activity of hLF, bLF and bLF hydrolysate was reported in an experimental mouse model (Bhimani et al., 1999). All the LF preparations demonstrated weak *in vitro* antibacterial activity while holo-LFs showed no activity. LF-treated mice (1 mg, i.v.) when injected with 10^6 staphylococci, showed 30-50% reduction in kidney infections, and viable bacterial counts in the kidney decreased 5- to 12-fold. The inhibitory effect was dose-dependent up to 1 mg LF. The LF preparations were effective when given 1 day prior to the bacterial challenge, after which there was no significant effect even at doses up to 5 mg. Apo- and Fe-saturated forms of hLF and bLF were all equally effective, while bLF hydrolysate was not protective. Different degrees of iron-saturation did not alter the *in vivo* antimicrobial property of either native LF preparation. Feeding mice with 2% bLF in drinking water also reduced the kidney infections by 40-60%, and viable bacterial counts, 5-12-fold.

Human LF was shown to be bactericidal *in vitro* for *Micrococcus luteus* but not for other *Micrococcus* species (*M. radiophilus, M. roseus* and *M. varians*) (de Lillo et al., 1997). A correlation between the binding of LF to the bacterial surface and the antimicrobial action was observed. Viability assays showed that ferric, but not ferrous, salts prevented binding and consequently *M. luteus* was not killed. The unsaturated form of LF showed a greater affinity than that of the iron-saturated molecule for lipomannan, a lipoglycan present on the cell wall of *M. luteus*, supporting the role for lipomannan as one of the possible binding sites of LF on *M. luteus*.

Custer and Hansen (1983) found that LF fragments could react with nitrite and cause inhibition of *Bacillus cereus* spore outgrowth. LF and lysozyme were shown to inhibit the growth of *Bacillus stearothermophilus* var. *calidolactis* spores (Carlsson et al., 1989). The growth of *Bacillus cereus* could be inhibited by LF and this effect could be reversed by the addition of iron (Sato et al., 1999). The growth inhibition was also reversed by the addition of erythrocytes and hemoglobin. *B. cereus* seems to use heme or heme-protein complex (hemoglobin-haptoglobin and hematin-albumin complexes) as iron sources in iron deficient conditions.

Oral administration of bLF with milk has been reported to inhibit various species of clostridia including *C. ramosum*, *C. paraputrificum* and *C. perfringens* in an experimental mouse model (Teraguchi et al., 1995).

TABLE 3. Inhibitory spectrum of hLF, bLF and LFcins against various bacteria.

Bacterial species	Form	Dose	Effect	Reference
Actino. actinomycetemcomitans	hLF	2 μM	Cidal (2-log reduction)	Kalmar & Arnold, 1988
Aeromonas hydrophila	bLF	0.1%	Adhesion-blockade (47%)	Paulsson et al., 1993
Bacillus cereus	bLFcin	6 μM	Cidal (4-log, 100%)	Hoek et al., 1997
Bacillus circulans	bLFcin	0.006%	Cidal (6-log, 100%)	Bellamy et al., 1992
Bacillus natto IFO3009	bLFcin	0.002%	Cidal (6-log, 100%)	Bellamy et al., 1992
Bacillus stearothermophilus	bLF	1:20	Stasis	Reiter & Oram, 1967
Bacillus subtilis	bLF	1:20	Stasis	Reiter & Oram, 1967
Bacillus subtilis ATCC6633	bLFcin	0.002%	Cidal (6-log, 100%)	Bellamy et al., 1992
Bifidobacterium longum	bLF	0.1%	Agglutination	Tomita et al., 1994
Corynebacterium diphtheriae	bLFcin	0.018%	Cidal (6-log, 100%)	Bellamy et al., 1992
Coryne. ammoniagenes	bLFcin	0.003%	Cidal (6-log, 100%)	Bellamy et al., 1992
Coryne. renale	bLFcin	0.001%	Cidal (6-log, 100%)	Bellamy et al., 1992
Clostridium innocuum	bLF	0.1%	Agglutination	Tomita et al., 1994
Clostridium perfringens	bLFcin	0.024%	Cidal (6-log, 100%)	Bellamy et al., 1992
Clostridium paraputrificum	bLFcin	0.003%	Cidal (6-log, 100%)	Bellamy et al., 1992
Enterococcus faecalis	bLFcin	0.06%	Cidal (6-log, 100%)	Bellamy et al., 1992
Escherichia coli E386	bLF	0.1%	Stasis (24-h, 100%)	Naidu et al., 1993
Escherichia coli	hLF	42 μM	Cidal (6-log reduction)	Arnold et al., 1980
Escherichia coli H10407	bLF	0.1%	Adhesion-blockade (50%)	Paulsson et al., 1993
Escherichia coli IID-861	bLFcin	10 μM	Cidal (3-log reduction)	Bellamy et al., 1992
Escherichia coli HB101	hLF	0.2%	Invasion-inhibition	Longhi et al., 1993
Escherichia coli CL99	bLF	20 μM	LPS release, OM damage	Yamauchi et al., 1993
Klebsiella pneumoniae	bLFcin	10 μM	Cidal (3-log reduction)	Bellamy et al., 1992
Lactobacillus casei	bLFcin	0.012%	Cidal (6-log, 100%)	Bellamy et al., 1992
Legionella pneumophila	hLF	0.03%	Cidal (4-log reduction)	Bortner et al., 1986
Listeria monocytogenes	bLFcin	10 μM	Cidal (4-log reduction)	Bellamy et al., 1992
L. monocytogenes NCTC7973	bLFcin	2 μM	Cidal (4-log, 100%)	Hoek et al., 1997
Micrococcus luteus	bLF	0.1%	Agglutination	Tomita et al., 1994
Proteus vulgaris JCM1668T	bLFcin	0.012%	Cidal (6-log, 100%)	Bellamy et al., 1992
Pseudomonas aeruginosa	hLF	42 μM	Cidal (7-log, 100%)	Arnold et al., 1980
Ps. aeruginosa IFO3446	bLFcin	10 μM	Cidal (3-log reduction)	Bellamy et al., 1992
Pseudomonas fluorescens	bLFcin	8 μM	Cidal (4-log, 100%)	Hoek et al., 1997
Salmonella abony	bLF	0.8%	Stasis (24-h, 100%)	Naidu & Arnold, 1994
Salmonella dublin	bLF	0.2%	Stasis (24-h, 100%)	Naidu & Arnold, 1994
Salmonella enteritidis	bLFcin	0.012%	Cidal (6-log, 100%)	Bellamy et al., 1992
Salmonella hartford	bLF	0.8%	Stasis (24-h, 100%)	Naidu & Arnold, 1994
Salmonella kentucky	bLF	0.2%	Stasis (24-h, 100%)	Naidu & Arnold, 1994
Salmonella panama	bLF	0.1%	Stasis (24-h, 100%)	Naidu & Arnold, 1994
Salmonella pullorum	bLF	0.2%	Stasis (24-h, 100%)	Naidu & Arnold, 1994
Salmonella rostock	bLF	0.2%	Stasis (24-h, 100%)	Naidu & Arnold, 1994
Salmonella salford	bLFcin	4 μM	Cidal (4-log, 100%)	Hoek et al., 1997
Salmonella montevideo	bLF	20 μM	LPS release, OM damage	Yamauchi et al., 1993
Salmonella thompson	bLF	0.1%	Stasis (24-h, 100%)	Naidu & Arnold, 1994
Salmonella typhimurium Rd	bLF	0.5%	Stasis (64%)	Naidu et al., 1993
Salm. typhimurium R10	bLF	0.1%	Adhesion-blockade (68%)	Paulsson et al., 1993
Salm. typhimurium SL696	bLF	20 μM	LPS release, OM damage	Yamauchi et al., 1993
Salmonella virchow	bLF	0.8%	Stasis (24-h, 100%)	Naidu & Arnold, 1994
Shigella flexneri	bLF	0.1%	Adhesion-blockade (30%)	Paulsson et al., 1993
Staphylococcus albus	bLF	0.5%	Stasis	Masson et al., 1966
Staphylococcus aureus	bLF	0.1%	Adhesion-blockade (54%)	Paulsson et al., 1993
Staph. aureus JCM2151	bLFcin	10 μM	Cidal (3-log reduction)	Bellamy et al., 1992
Staphylococcus epidermidis	bLFcin	0.006%	Cidal (6-log, 100%)	Bellamy et al., 1992
Staphylococcus haemolyticus	bLFcin	0.001%	Cidal (6-log, 100%)	Bellamy et al., 1992
Staphylococcus hominis	bLFcin	0.003%	Cidal (6-log, 100%)	Bellamy et al., 1992

Bacterial species	Form	Dose	Effect	Reference
Streptococcus bovis	bLFcin	0.006%	Cidal (6-log, 100%)	Bellamy et al., 1992
Streptococcus cremoris	bLFcin	0.003%	Cidal (6-log, 100%)	Bellamy et al., 1992
Streptococcus lactis	bLFcin	0.003%	Cidal (6-log, 100%)	Bellamy et al., 1992
Streptococcus mitior	hLF	42 μM	Cidal (6-log, 100%)	Arnold et al., 1980
Streptococcus mutans AHT	hLF	0.17%	Cidal (7-log, 100%)	Arnold et al., 1977
Strep. mutans LM-7	hLF	42 μM	Cidal (8-log, 100%)	Arnold et al., 1980
Strep. mutans ATCC25175	hLF	0.01%	Agglutination	Soukka et al., 1993
Streptococcus pneumoniae	hLF	42 μM	Cidal (7-log, 100%)	Arnold et al., 1980
Streptococcus salivarius	hLF	83 μM	Cidal (7-log, 100%)	Arnold et al., 1980
Streptococcus thermophilus	bLFcin	0.003%	Cidal (6-log, 100%)	Bellamy et al., 1992
Vibrio cholerae 569B	hLF	0.33%	Cidal (7-log, 100%)	Arnold et al., 1977

Groenink and co-workers (1999) reported a potent antimicrobial activity of synthetic cationic peptides derived from the N-terminal domain that comprises an amphipathic a-helix in hLF (hLF 18-31 and hLF 20-38) and bLF (bLF 17-30 and bLF 19-37). Peptide bLF 17-30, containing the largest number of positively charged amino acids, elicited the highest inhibitory spectrum against both Gram-positive and Gram-negative bacteria.

*2. **Gram-negative bacteria.*** Many studies have shown the antimicrobial activity of LF against Gram-negative bacteria, *E. coli*, in particular. Various antimicrobial effects of LF were demonstrated against *E.coli* and different mechanisms were postulated to elucidate these effects. LF elicits a bacteriostatic effect on *E. coli*. Based on the time required for *E.coli* O111 to reach one-half maximal cell density, Stuart and co-workers (1984) indicated that the *in vitro* effects of LF on the growth of *E. coli* was kinetic rather than bacteriostatic. Compared to a control, added apo-LF (0.25-1.0 mg/ml) produced only a delay effect indicating that these concentrations are probably within the sub-inhibitory concentration range. The kinetic delay effect of apo-LF also increased steadily in the presence of Zn^{2+} and Cu^{2+} cations. Cu^{2+}, Zn^{2+} and nitrilotriacetate did not affect the growth rate of this organism in the absence of LF compared to the control. These studies indicate that the mechanism by which LF alters the bacterial growth of *E. coli* O111 is more complex than simple iron deprivation.

Rainard (1987) reported the antibacterial activity of milk against a virulent strain of *E. coli* using milk fractions from normal or inflamed glands. Whey obtained from mastitis milk exhibited either bactericidal or bacteriostatic activities, depending on whether bacteria were enumerated by the pour plate technique or by surface plating onto sheep blood agar. The cidal activity, however, was not due to LF, even when assayed in distilled water. Milk whey ultra-filtrate was used to assay the ability of normal and mastitis milk to support the antibacterial activities of LF against *E. coli*. The addition of purified LF to ultra-filtrate from mastitis whey resulted in bacteriostasis, whereas LF was without effect in ultra-filtrate from normal whey. It was suggested that LF could inhibit the growth of LF-sensitive bacteria during mastitis depending on plasma exudation during mastitis. Dionysius et al. (1993) reported that the antibacterial activity of apo-LF depends on bacterial inoculum size and the addition of holo-LF or cytochrome-C could diminish the effect. Furthermore, the resistance to inhibition by LF was not related to the production of bacterial siderophores in *E.coli*.

Ellison et al. (1988) hypothesized that the iron-binding proteins could affect the Gram-negative outer membrane in a manner similar to that of the chelator EDTA. Further, both the whole protein and a cationic N-terminus peptide fragment directly damage the outer membrane of Gram-negative bacteria suggesting a mechanism for the supplemental effects. Several groups have also shown that LF could synergistically interact with immunoglobins, complement, and neutrophil cationic proteins against Gram-negative bacteria.

Klebanoff and Waltersdorf (1990) found that Fe^{2+} and apo-LF could generate hydroxy radicals via an H_2O_2 intermediate with toxicity to *E.coli*, and hypothesized that such a mechanism could possibly contribute to the microbicidal activity of phagocytes.

LF binds to surface structures expressed in *E. coli* K-12 strains grown under iron limitation (Visca et al., 1990). Both apo and holo forms of LF yielded a maximum of 1.6 X 10^5 bound molecules/*E. coli* K-12 cell. The amount of LF bound was independent of the expression of iron-regulated outer membrane proteins. However, LF did not bind to *E. coli* clinical isolates. Apo-LF (500 µg/ml) in a chemically defined medium inhibited the growth of *E. coli* K-12 strains but not of clinical isolates. These findings suggested that the antibacterial activity of the protein could be associated to its binding to the cell surface. Enterotoxigenic strains demonstrate higher LF interaction than enteropathogenic, enteroinvasive, enterohemorrhagic strains or normal intestinal *E. coli* isolates (Naidu et al., 1991). Also the enteropathogenic strains belonging to serotypes 044 and 0127 demonstrate higher LF binding compared to 026, 055, 0111, 0119 and 0126 serotypes. No significant differences in the degree of hLF or bLF binding were noticed between aerobactin-producing and non-producing strains. In later studies, Naidu and co-workers have identified and characterized porins in the outer membrane of Gram-negative bacteria as the specific receptors for LF interaction (Tigyi et al., 1992; Naidu et al., 1993; Erdei et al., 1994)

Using an *in vitro* model, Gutteberg and co-workers (1990) reported the early effect of *E. coli*, *Streptococcus agalactiae* (group B streptococci, GBS) and recombinant tumor necrosis factor alpha (TNF) on the release of LF and the generation of interleukin-1 (IL-1) due to *E. coli*, using heparinized whole blood from healthy full-term newborns. In a final concentration of 10^7 per ml both bacteria increased the release of LF markedly. The response to *E. coli* was immediate. GBS was a less potent stimulant than *E. coli* and the response was only apparent after 20 minutes. TNF in a concentration of 10 ng/ml as well as 1 ng/ml increased the release of LF significantly, whereas a concentration of 0.1 ng/ml had no effect. Whole blood incubated with different preparations of LF for 20 minutes did not increase the LF levels. No significant changes in IL-1 levels were observed. LF had bacteriostatic but no bactericidal effect on GBS and *Streptococcus mutans*.

Payne et al. (1990) demonstrated that apo-bLF had bacteriostatic activity against four strains of *L. monocytogenes* and an *E.coli* at concentrations of 15 to 30 mg/ml, in UHT milk. At 2.5 mg/ml the compound has no activity against *S. typhimurium, P. fluorescens* and limited activity against *E.coli* O157:H7 or *L. monocytogenes* VPHI (Payne et al., 1994).

Human LF and free secretory component (fSC) were shown to inhibit the hemagglutination induced by *E. coli* CFA1+ (Giugliano et al., 1995). The lowest concentrations of purified fSC and hLF to inhibit the hemagglutination by *E. coli* strain TR50/3 CFA1+ were 0.06 mg/ml and 0.1 mg/ml, respectively.

TABLE 4: Lactoferrin – Antiviral effects

Viral pathogen	Antiviral effect	Reference
Spleen focus forming virus (SFFV)	Decreases multiplication	Hangoc et al., 1987
Human influenza virus	Inhibits viral hemagglutination	Kawasaki et al., 1993
Human cytomegalovirus (HCMV)	Inhibits infection & replication	Hasegawa et al., 1994
	Inhibits MT4 cell cytopathy	Harmsen et al., 1995
Human herpes simplex virus (HSV-1)	Inhibits adsorption & penetration	Hasegawa et al., 1994
	Prevents plaque formation	Fujihara & Hayashi, 1995
Human immunodeficiency virus (HIV)	Inhibits MT4 cytopathic effect	Harmsen et al., 1995
	Inhibits vero cell cytopathy	Marchetti et al., 1996
Feline immunodeficiency virus (FIV)	Effects clinical outcome	Sato et al., 1996
Respiratory syncytial virus	Inhibits viral multiplication	Grover et al., 1997
Hepatitis C virus (HCV)	Binds E1 and E2 envelope proteins	Yi et al., 1997
Rotavirus	Inhibits HT-29 cell infection	Superti et al., 1997

The antimicrobial activities of bLF, and bLFcin against four clinical isolates of enterohemorrhagic *E. coli* O157:H7 were reported (Shin et al., 1998). The MICs against these isolates were 3 mg/ml for bLF, and 8-10 μg/ml for bLFcin in 1% Bacto-peptone broth. Transmission electron microscopy findings suggested that bLFcin acts on the bacterial surface and affects cytoplasmic contents. Furthermore, bLFcin affected the levels of verotoxins in the culture supernatant fluid of an *E. coli* O157:H7 strain.

The antimicrobial effect of LF against *Salmonella typhimurium* was tested by measuring conductance changes in the cultivation media by using a Malthus-AT system (Naidu et al., 1993). Conductance measurement studies revealed that the low-LF-binding strain 395MS with smooth LPS was relatively insusceptible to LF, while the high-LF-binding mutant Rd with rough LPS was more susceptible to LF suggesting an LPS shielding of antimicrobial effect. Later studies have led to the identification of porins as LF-binding outer membrane proteins in various species of *Salmonellae* (Naidu & Arnold, 1994).

Antimicrobial effects of LF against various Gram-negative bacterial pathogens, including *Aeromonas hydrophila, Yersinia enterocolitica, Campylobacter jejuni, Helicobacteri pylori, Pseudomonas aeruginosa, Vibrio sp.*, have also been reported (Arnold et al., 1977; Paulsson et al., 1993; Tomita et al., 1994).

B. Antiviral activity

LF demonstrates a broad-spectrum antiviral activity against both DNA and RNA viruses (*TABLE 4*). The ability of LF to interact with nucleic acids as well as its capacity to bind eucaryotic cells and prevent viral adhesion seem to be the possible antiviral mechanisms.

Abramson et al. (1984) suggested that the depressed chemotactic activity of PMNL infected with influenza virus could be due to changes occurring at the plasma membrane. Virus-treated PMNL stimulated with FMLP or *Staphylococcus aureus* exhibited a marked decrease for LF released into phagosomes, onto the cells' outer membrane, and into the extracellular medium as compared to control cells. Baynes et al. (1988) with the use of an immunoperoxidase stain for LF, showed that neutrophils in viral illness have reduced LF content compared to normal subjects. The authors suggested an acquired defect of neutrophil LF synthesis in viral infection. The LF levels in parotid saliva from

individuals with different clinical stages of human immunodeficiency virus (HIV) infection were significantly decreased in parallel with their markedly reduced parotid secretory IgA output. This combined deficiency of parotid LF and secretory IgA may well contribute to the frequent oral infections seen in subjects with HIV infection (Muller et al, 1992).

Purified holo-hLF and recombinant murine IL-3 were assessed *in vivo* for their effects on replication of Spleen Focus Forming Viruses (SFFV) in spleens of DBA/2 mice injected with the polycythemia-inducing strain of the Friend Virus Complex (FVC) (Hangoc et al, 1987). LF and IL-3, inoculated 2 hr prior to the administration of the polycythemia-inducing strain of the FVC, respectively decreased and increased the replication of SFFV in mice as assessed by the spleen focus forming unit assay in primary and secondary DBA/2 mice. Since virus infectivity is associated with the DNA synthetic phase of the cell cycle and it has been shown that LF decreases and IL-3 increases the percent of hematopoietic progenitor cells in S-phase *in vivo*, the results suggest that the opposing actions of LF and IL-3 on replication of SFFV may reflect the actions of these molecules on cycling of the target cells for SFFV.

Human LF and bLF inhibit the infection of tissue culture cells with human cytomegalovirus (HCMV) and human herpes simplex virus-1 (HSV-1) (Hasegawa et al, 1994). The addition of LF inhibited both *in vitro* infection and replication of HCMV and HSV-1 in human embryo lung host cells. The maximum inhibition by more than six exponential of ID_{50} for HCMV and four exponential for HSV-1 was obtained at a concentration in a range from 0.5 to 1 mg of LF per ml of medium. The antiviral activity of LF was associated with its protein moiety, but not with its iron molecule or sialic acid. Furthermore, LF prevented virus adsorption and/or penetration into host cells, indicating an effect on the early events of virus infection. Preincubation of host cells with LF for 5 to 10 min was sufficient to prevent HCMV infection, even when LF was removed after addition of virus. These results suggest that LF possesses a potent antiviral activity and may be useful in preventing HCMV and HSV-1 infection in humans.

Native and chemically derivatized proteins purified from serum and milk were assayed *in vitro* to assess their inhibiting capacity on the cytopathic effect of human immunodeficiency virus (HIV)-1 and human cytomegalovirus (HCMV) on MT4 cells and fibroblasts, respectively (Harmsen et al, 1995). Only native and conformationally intact LF from bovine or human milk, colostrum, or serum could completely block HCMV infection (IC_{50} = 35-100 μg/mL). Moreover, native LF also inhibited the HIV-1-induced-cytopathic effect (IC_{50} = 40 μg/mL). When negatively charged groups were added to LF by succinylation, there was a four-fold stronger antiviral effect on HIV-1, but the antiviral potency for HCMV infection decreased. LF likely exerts its effect at the level of virus adsorption or penetration (or both), because after HCMV penetrated fibroblasts, the ongoing infection could not be further inhibited. A number of native and modified milk proteins from bovine or human sources were analyzed for their inhibitory effects on human immunodeficiency virus type 1 (HIV-1) and HIV-2 *in vitro* in an MT4 cell test system (Swart et al, 1996). The proteins investigated were LF, α-lactalbumin, β-lactoglobulin A, and β-lactoglobulin B. By acylation of the amino function of the lysine residues in the proteins, using anhydrides of succinic acid or cis-aconitic acid, protein derivatives were obtained that all showed a strong antiviral activity against HIV type 1 and/or 2. The *in vitro* IC_{50} values of the aconitylated proteins were in the concentration range of 0.3 to 3

nM. Succinylation or aconitylation of α-lactalbumin and β-lactoglobulin A/B also produced strong anti-HIV-2 activity with IC_{50} values on the order 500 to 3000 nM. All compounds showed virtually no cytotoxicity at the concentration used. Peptide-scanning studies indicated that the native LF as well as the charged modified proteins strongly binds to the V3 loop of the *gp120* envelope protein, with K_a values in the same concentration range as the above-mentioned IC_{50}. Therefore, shielding of this domain, resulting in inhibition of virus-cell fusion and entry of the virus into MT4 cells, may be the likely underlying mechanism of antiviral action.

LF prevents herpes simplex virus type-1 (HSV-1) plaque formation in Vero cell monolayer (Fujihara & Hayashi, 1995). Topical administration of 1% LF prior to the virus inoculation suppressed infection on ocular tissue, however it did not inhibit propagation of the virus in the mouse cornea. Marchetti et al. (1996) reported that both hLF and bLF are potent inhibitors of HSV-1 infection, the concentrations required to inhibit the cytopathic effect in Vero cells by 50% being 1.41 μM and 0.12 μM, respectively. Human LF and bLF exerted their activity through the inhibition of adsorption of virions to the cells independently of their iron withholding property showing similar activity in the apo- and iron-saturated form. The binding of [^{35}S]methionine-labelled HSV-1 particles to Vero cells was strongly inhibited when bLF was added during the attachment step. Bovine LF interacts with both Vero cell surfaces and HSV-1 particles, suggesting that the hindrance of cellular receptors and/or of viral attachment proteins may be involved in its antiviral mechanism.

The effect of LF on the growth of rotavirus and respiratory syncytial virus in cell culture was investigated (Grover et al., 1997). LF inhibited the growth of respiratory syncytial virus at a concentration tenfold lower than that normally present in human milk. The ability of LF to inhibit influenza virus hemagglutination was also reported (Kawasaki et al., 1993).

Hepatitis C virus (HCV) has two envelope proteins, E1 and E2, which form a hetero-oligomer. During dissection of interacting regions of HCV E1 and E2, Yi et al. (1997) found the presence of an interfering compound or compounds in skim milk identified as LF. The bindings of LF to HCV envelope proteins *in vitro* were confirmed by Western blotting and by the pull-down assay, with immuno-precipitated LF-bound protein A resin. Direct interaction between E2 and LF was proved *in vivo*, since anti-hLF antibody efficiently immuno-precipitated with secreted and intracellular forms of the E2 protein. The N-terminal loop of LF, the region important for the antibacterial activity, has only a little role in the binding ability to HCV E2 but affected the secretion or stability of LF. Taken together, these results indicate the specific interaction between LF and HCV envelope proteins *in vivo* and *in vitro*.

Different milk proteins were analyzed for their inhibitory effect on either rotavirus-mediated agglutination of human erythrocytes or rotavirus infection of the human enterocyte-like cell line HT-29 (Superti et al., 1997). Apo- and Fe-LF inhibited the replication of rotavirus in a dose-dependent manner, apo-LF being the most active. It was shown that apo-LF hinders virus attachment to cell receptors since it is able to bind the viral particles and to prevent both rotavirus hemagglutination and viral binding to susceptible cells. Moreover, LF markedly inhibited rotavirus antigen synthesis and yield in HT-29 cells when added during the viral adsorption step or when it was present in the first hours of infection, suggesting that this protein interferes with the early phases of rotavirus infection.

C. Antifungal activity

LF and lysozyme (muramidase), either singly or in combination, are fungicidal in nature and their combined activity is synergistic. Samaranayake and co-workers (1997) examined 20 oral isolates of *Candida krusei* and 5 isolates of *Candida albicans* for their susceptibility to human apo-LF and lysozyme, either singly or in combination, using an *in vitro* assay system. The two species exhibited significant interspecies differences in susceptibility to LF, but not for lysozyme; *C. krusei* was more sensitive to LF (1.4 times) than *C. albicans*. Both species revealed significant intraspecies differences in their susceptibility to lysozyme, but not for LF. No synergistic antifungal activity of the two proteins on either *Candida* species was noted.

LF could inhibit the growth of *C.albicans* in the absence of PMNL, and anti-LF antibodies reversed both this inhibition and the PMNL activation by MP-F2, GM-CSF, and LPS. Furthermore, PMNL may be activated by relevant candidal mannoproteins, and release of LF may add to other antimicrobial mechanisms of PMNL for the control of candidal infections (Palma et al., 1992).

Candida albicans was found highly susceptible to inhibition and inactivation by bLFcin, a peptide produced by enzymatic cleavage of bLF (Bellamy et al., 1993). Effective concentrations of the peptide varied within the range of 18 to 150 μg/ml depending on the strain and the culture medium used. Its effect was lethal, causing a rapid loss of colony-forming capability. ^{14}C-labeled bLFcin bound to *C. albicans* and the rate of binding appeared to be consistent with the rate of killing induced by the peptide. The extent of binding was diminished in the presence of Mg^{2+} or Ca^{2+} ions, which acted to reduce its anticandidal effectiveness. Binding occurred optimally at pH 6.0 and killing was maximal near the same pH. Such evidence suggests the lethal effect of bLFcin results from its direct interaction with the cell surface. Cells exposed to bLFcin exhibited profound ultrastructural damage that appeared to reflect its induction of an autolytic response.

D. Antiparasitic activity

LF could elicit defense against parasitic infections by phagocytic activity in the destruction of amastigotes, an intracellular parasitic form of *Trypanosoma cruzi* in macrophages. The effect of bLF on the intracellular growth *Toxoplasma gondii* parasites was examined in murine macrophage and embryonic cells (Tanaka et al., 1996). Co-cultures of host cells with the parasites were supplemented with LF, apo-LF, holo-LF or TF in the culture media for varying periods. The growth activity of intracellular parasites in the host cells was determined by the measurement of selective incorporation of ^{3}H-uracil. Supplement of LF had no effect on the penetration activity of the parasites, while development of intracellular parasites was inhibited linearly in concentration of LF. Supplement of apo-LF and holo-LF, but not TF showed similar effects. These suggest that LF induce the inhibitory effects on the development of intracellular parasites. Pretreatment of LF to the macrophages, however, did not show any inhibitory effects, whereas mouse embryonic cells preincubated with LF suppressed the intracellular growth. Thus, the action of LF to macrophages would be different from that of mouse embryonic cells.

The trophozoites of *Giardia lamblia* could be killed by nonimmune human milk in a time- and concentration-dependent manner. Removal of greater than 99% of the S-IgA from milk did not decrease its *Giardia*-cidal activity. Thus, the killing was not anti-

body dependent. Studies by Gillin and co-workers (1983) showed that in the presence of milk, trophozoites lost motility, swelled, and lysed. The *Giardia*-cidal activity may be specific to human milk, since unheated cow and goat milk were virtually devoid of activity. Human and bLFs and their derived N-terminal peptides were giardicidal *in vitro* (Turchany et al., 1995). Fe^{3+}, but not Fe^{2+}, protected trophozoites from both native LF and peptides, although the latter lack iron-binding sites. Other divalent metal ions protected only against native LF. Log-phase cells were more resistant to killing than stationary-phase cells. These studies suggest that LF, especially in the form of the N-terminal peptides, may be an important nonimmune component of host mucosal defenses against *Giardia lamblia*.

VII. INFLUENCING FACTORS

A. Citrate / bicarbonate ratio

The requirement of bicarbonate for formation of the red complex of LF was first pointed out by Blanc and Isliker (1963). In a later study, Masson and Heremans (1968) clearly established the involvement of bicarbonate in the metal-combining properties of LF. It was found that one molecule of bicarbonate (carbon dioxide) is taken up per atom of iron or copper during the formation of the colored LF-metal complex. It was observed that the color development proceeded very slowly when a gas-free sample of apo-LF was exposed to air in the presence of copper ions, whereas it took place instantly when bicarbonate was added. This suggested that the carbon dioxide from the air first had to become converted to bicarbonate ions before its participation in the reaction.

Fresh human and bovine milk are bacteriostatic *in vitro* for certain (milk-sensitive) strains of *E. coli*. Dolby et al. (1977) reported that the addition of bicarbonate to the test system could potentiate the bacteriostasis which also results in the inhibition of milk-resistant strains. The concentration of bicarbonate needed for such an effect is lower for human milk than for cow milk and is reduced even further by the addition of more LF. *In vivo* studies with infants and data deduced from the ratio of milk-sensitive to milk-resistant strains of *E. coli* isolated from fecal samples suggested that the neonatal intestinal secretions may contribute to the bacteriostatic activity of their feeds so that (i) in fully breastfed babies all strains of *E. coli* are inhibited to the same extent; there is no selection on the basis of milk sensitivity and equal numbers of strains resistant and sensitive to milk are found in the feces; (ii) in fully bottle-fed babies *E. coli* is not inhibited since the milk is non-bacteriostatic and again there is no selection; (iii) in babies fed at the breast but bottle-milk supplemented, only milk-sensitive strains are inhibited; milk-resistant strains are not, and preferentially colonized the large intestines.

Undiluted dialyzed milk was not inhibitory because of its low content of LF and the lack of inhibition in undiluted whey is due to the high concentration of citrate in colostral whey (and milk) (Reiter et al., 1975). It was suggested that citrate competes with the iron-binding proteins for iron and makes it available to the bacteria. Addition of bicarbonate, which is required for the binding of iron by LF, can overcome the effect of citrate. Coliform bacteria respond to low-iron environments by production of iron-sequestering agents that compete effectively with apo-LF for free iron. Addition of apo-LF plus citrate could reverse growth inhibition. The molar ratio (citrate to apo-LF) is more important than the absolute concentration of either component. Bishop et al. (1976) found that a ratio of

75 resulted in 50% growth inhibition, whereas ratios of 300 and greater resulted in less than 10% growth inhibition. These results suggest that the ratio of citrate to LF is important in the nonspecific protection of bovine udder. Inhibition of *E.coli* by apo-LF (10 mg/ml) was abolished by addition of ferric iron to the assay system, indicating an iron-dependent nature of apo-LF induced inhibition of bacteria (Nonnecke & Smith, 1984a). Bicarbonate supplementation of the growth system containing apo-LF (1 mg/ml) increased inhibition of three coliform strains by apo-LF. Addition of increasing concentrations of citrate (2.0 mg/ml) to an assay system containing apo-LF (5 mg/ml) resulted in a concomitant reduction of growth inhibition of three coliform strains. These data further indicate a potential relationship between the molar ratio of citrate to LF of the lacteal secretion and its capacity to inhibit coliform strains associated with mastitis. Changes in pH, concentration of serum albumin, immunoglobulin G, citrate, LF, and number of leukocytes in secretions were typical of milk from glands undergoing physiological transitions. Whey from different glands of the same cow differ markedly in their capacity to inhibit growth of coliforms. Bacteriostatic activity of whey increases markedly during the dry period and reaches maximal in whey collected on day-15 of the dry period and at parturition in the subsequent lactation (Nonnecke & Smith, 1984b).

Griffiths and Humphreys (1977) conducted a series of experiments to elucidate the importance of bicarbonate and milieu pH in the bacteriostatic activity of LF. At pH 7.4 and in the presence of bicarbonate, human milk and bovine colostrum inhibit the growth of *E. coli* O111. Adding sufficient iron to saturate the iron-binding capacity of the LF present in the milk or colostrum prevents bacteriostasis. At pH 6.8, neither milk nor colostrum could inhibit *E. coli* 0111. Adjusting the pH to 7.4 with bicarbonate resulted in the development of bacteriostatic activity. Adjusting the pH to 7.4 with NaOH was ineffective. Dialyzed colostrum and milk inhibited bacterial growth at pH 6.8 in the absence of added bicarbonate; addition of citrate or iron abolished bacteriostasis. The chromatographic elution profile of tyrosyl-tRNA from iron-replete *E. coli* differs significantly from that of tyrosyl-tRNA from iron-deficient organisms. Examination of the elution profile tyrosyl-tRNA from *E. coli* 0111 growing in colostrum without added bicarbonate showed that such bacteria were fully replete in iron. The nature of the elution profile of tyrosyl-tRNA also showed that iron was freely available to the bacteria when citrate was added to dialyzed colostrum but not available in its absence, even at pH 6.8. These data support the idea that the bacteriostatic action of milk and colostrum, due to the combined action of antibody and LF, depends on the addition of bicarbonate to counteract the iron-mobilizing effect of the citrate normally present in these secretions.

Thomas and Fell (1985) reported that in lactating cows hormones [oxytocin or ACTH (Synacthen)] affect the citrate and LF concentrations in the direction that would improve the antibacterial properties of milk, but that this was accompanied by adverse effects on milk secretion. However, the extent of the change was not sufficient to produce inhibition of coliform bacteria.

B. Milieu pH

In order to apply functionally active bLF to food products, the effect of pH on the heat stability of bLF was studied (Saito et al., 1994). Bovine LF was easily denatured to an insoluble state by heat treatment under neutral or alkaline conditions, above pH 6.0. In contrast, it remained soluble after heat treatment under acidic conditions at pH 2.0 to 5.0, and the HPLC pattern of LF heat-treated at pH 4.0 at 100°C for 5 min was the same as that of native bLF. Bovine LF was thermostable at pH 4.0, and could be pasteurized or

sterilized without any significant loss of its physicochemical properties. Bovine LF was hydrolyzed by heat treatment at pH 2.0 to 3.0 at above 100°C, and its iron binding capacity and antigenicity were lost.

Acceleration of the autoxidation of Fe(II) by apoTF or apo-LF at acid pH is indicated by the disappearance of Fe(II), the uptake of oxygen, and the binding of iron to TF or LF. The product(s) formed oxidize iodide to an iodinating species and are bactericidal to *E. coli* (Klebanoff & Waltersdorph, 1990) Toxicity to *E. coli* by $FeSO_4$ (10 μM) and human apo-TF (100 μg/ml) or apo-hLF (25 μg/ml) was optimal at acid pH (4.5-5.0) and with log-phase organisms. Both the iodinating and bactericidal activities were inhibited by catalase and the hydroxyl radical (OH.) scavenger mannitol, whereas superoxide dismutase was ineffective. NaCl at 0.1 M inhibited bactericidal activity, but had little or no effect on iodination. Iodide increased the bactericidal activity of Fe(II) and apoTF or apo-LF. The formation of OH.was suggested by the formation of the OH.spin-trap adduct (5,5-dimethyl-1-pyroline N-oxide [DMPO]/OH)., with the spin trap DMPO and the formation of the methyl radical adduct on the further addition of dimethyl sulfoxide. (DMPO/OH).formation was inhibited by catalase, whereas superoxide dismutase had little or no effect. These data suggest that Fe(II) and apoTF or apo-LF can generate OH. via an H_2O_2 intermediate with toxicity to microorganisms, and raise the possibility that such a mechanism may contribute to the microbicidal activity of phagocytes.

Salamah and al-Obaidi (1995b) studied the effect of pH, temperature, magnesium and calcium on the bactericidal activity of LF and TF against *Yersinia pseudotuberculosis*. The bactericidal activity of LF was higher at acid pH, whereas the bactericidal activity of TF was higher at alkaline pH. Neither was efficient at 4°, 15°, and 25°C, but both were effective at 37°C. LF, but not TF, was very efficient at 42°C. The activity of both were time and concentration dependent. Calcium did not effect their activity up to 60 mM, whereas magnesium reduced the activity of LF only.

LF release from the secondary granules of activated PMNs is markedly lower at pH 7.2 than at pH 6.7 or 8.2 (Leblebicioglu et al., 1996). Moreover, phagocytosis of opsonized bacteria is lower at pH 7.2 than at pH 7.7. In addition to these effects on functional activation, extracellular pH influences the magnitude of intracellular Ca^{2+} mobilization. These findings suggest that the pH of an inflammatory milieu can selectively influence PMN activation, thereby altering the balance between bacteria and the host response.

C. Proteases

Holo-bLF is more resistant to proteolysis than the apo-form (Brock et al., 1976; 1978). In the trypsin digests of bLF, up to five different fragments with molecular weights ranging from 25-kDa to 53-kDa were detected, with no obvious qualitative difference between digests of holo and apo forms. The susceptibility of apo-bLF to tryptic digestion was only slightly reduced when the protein was complexed with β-lactoglobulin, suggesting that complex-formation is not a mechanism for protecting LF against intestinal degradation.

The susceptibility of hLF and bLF to digestion by trypsin and chymotrypsin has been compared (Brines & Brock, 1983). Neither enzyme had much effect on the hLF-mediated antimicrobial activity of human milk, and the iron binding capacity of hLF in the milk was only slightly reduced. Both enzymes had only a slight effect on the iron bind-

ing capacity of purified hLF. In contrast, trypsin destroyed the antimicrobial activity of bovine colostrum, and, in line with earlier studies, appreciably reduced the iron binding capacity of both colostrum and purified apo-bLF. Holo-bLF was more resistant to digestion. The unusual resistance of apo-hLF to proteolysis may reflect an evolutionary development designed to permit its survival in the gut of the infant.

Proteolytic hydrolysis of milk proteins in the mammary gland during involution could influence the bio-functionality of LF and other innate defense factors. Aslam and Hurley (1997) examined the activities of plasmin, plasminogen, and plasminogen activator on proteolysis of LF in mammary gland secretions collected during involution. Activities of plasmin, plasminogen, and plasminogen activator were significantly higher on day-7, -14, and -21 of involution than were those on day-7 postcalving. Protein fragments resulting from hydrolysis were detected by SDS-PAGE in samples collected on day-7, -14, and -21 of involution, but few protein fragments were observed in samples collected on day-7 postcalving when plasmin activity was low. Immunoblot analysis showed that a number of peptides observed during involution were generated from α-s-casein (CN), β-CN, κ-CN, or LF. The appearance of peptides from proteins of mammary secretions during early involution was generally correlated with increased plasmin activity. Elevated plasmin activity during mammary involution may be primarily responsible for the observed concurrent hydrolysis of milk proteins in mammary secretions.

The ability of periodontal pathogens *Porphyromonas gingivalis, Prevotella intermedia* and *Prevotella nigrescens* to degrade LF was reported (de Lillo et al., 1996). Strains of *P. gingivalis* completely degraded LF *in vitro*, whereas *P. intermedia* and *P. nigrescens* showed only partial degradation. It was suggested that LF binds to a high-affinity receptor on all these bacteria and, particularly in the case of *P. gingivalis*, is then degraded by cell-associated proteases. This property may provide protection to the cell against the effects of LF in periodontal sites and so is a possible virulence factor in disease. However, there was no association between the ability to degrade LF and whether the strains had originated from healthy or diseased oral sites.

Bacterium- and neutrophil-derived proteases have been suggested to contribute to tissue injury at sites of *Pseudomonas aeruginosa* infection. *Pseudomonas* elastase cleavage of LF and TF enhances *in vitro* iron removal from these proteins by the *P. aeruginosa* siderophore pyoverdin. Britigan et al. (1993) detected TF and LF cleavage products in bronchoalveolar lavage (BAL) samples from 21 of 22 and 20 of 21 cystic fibrosis (CF) patients, respectively. Three of eleven and two of nine BAL samples from individuals with other forms of chronic inflammatory lung disease had TF and LF cleavage products, respectively. Each patient in whom such products were detected was also infected with *P. aeruginosa*. No such products were detected in normal individuals. These data provide evidence that *P. aeruginosa*- and/or human-derived protease cleavage of TF and LF occurs *in vivo* in the airways of individuals with CF and other forms of chronic lung disease, suggesting that this process could contribute to *P. aeruginosa*-associated lung injury in these patients.

The effect of *Vibrio cholerae* non-O1 protease on LF was studied in relation to bacterial virulence mechanism (Toma et al., 1996). The proteins treated with the protease were analyzed by SDS-PAGE. The protease has cleaved LF into two fragments of 50-kDa and 34-kDa. The N-terminal amino acid sequencing of these fragments revealed that the cleavage site was near the hinge region, between *serine 420* and *serine 421*. This cleav-

age could affect the transition from open to closed configuration, which is involved in iron binding and release. However, the anti-bacterial activity of LF was not affected by protease treatment.

D. Microbial iron acquisition

Iron is essential to all microorganisms. The low concentration of free iron in body fluids creates bacteriostatic conditions for many microorganisms and is therefore an important defense factor of the body against invading bacteria. Iron-binding proteins, such as LF, TF, and ferritin, play a central role in human ferrokinetics. These iron-binding proteins also participate in the process of decreasing iron availability for the microorganisms. LF and TF restrict the amount of ionic iron available in body fluids to 10^{-18} M (Bullen, 1981). They do so by decreasing iron re-utilization. Anemia of inflammation (previously called anemia of chronic disease) is seen in the setting of infectious, inflammatory, and neoplastic diseases. It results, in part, from changes in the intracellular metabolism of iron. Alterations of iron physiology seen in many clinical circumstances make excess iron available to microorganisms, thus enhancing their pathogenicity. Understanding the molecular basis of iron withholding by the human host, both in the absence of and during infection, and that of iron acquisition by microorganisms may provide us with new and innovative antimicrobial agents and vaccines.

Pathogenic bacteria have developed several mechanisms for acquiring iron from the host (see reviews: Otto et al., 1992; Crosa, 1989). Siderophore-mediated iron uptake involves the synthesis of low molecular weight iron chelators called siderophores which compete with the host iron-binding glycoproteins LF and TF for iron. Other ways to induce iron uptake, without the mediation of siderophores, are the possession of outer membrane protein receptors that actually recognize the complex of TF or LF with iron, resulting in the internalization of this metal, and the use of heme-compounds released into the circulation after lysis of erythrocytes.

Rogers and Synge (1978) reported that enterochelin, an iron transporting compound of *E.coli* could abolish the bacteriostatic effect of human milk. The bacteriostatic phase in human milk could be abolished by adding sufficient iron to saturate the LF in human milk, and also by adding supernatant from a 24-h milk culture or by adding enterobactin, an enterobacterial iron chelator (Brock et al., 1983). Growth in the presence of enterobactin was even more rapid than in the presence of excess iron. Partial loss of bacteriostatic activity could be achieved by absorbing the milk with bacterial antigens, but no clear correlation with removal of antibodies to O, K, or H antigens was apparent.

Many strains of *E. coli* are able to synthesize two siderophores, aerobactin and enterochelin. Although aerobactin has a dramatically lower affinity for iron than enterochelin, it has been shown to provide a significant selective advantage for bacterial growth in conditions of iron limitation, such as in the body fluids and tissues. Differential regulation of the genetic determinants of the two siderophores resulted in preferential induction of the aerobactin system in the presence of unsaturated levels of TF and LF (Williams & Carbonetti, 1986).

Pathogenic *Neisseriae* have a repertoire of high-affinity iron uptake systems to facilitate acquisition of this essential element in the human host. They possess surface receptor proteins that directly bind the extracellular host iron-binding proteins TF and LF (see review: Schryver & Stojiljkovic, 1999). Alternatively, they have siderophore recep-

tors capable of scavenging iron when exogenous siderophores are present. Released intracellular heme iron present in the form of hemoglobin, hemoglobin-haptoglobin or free heme can be used directly as a source of iron for growth through direct binding by specific surface receptors. Although these receptors may vary in complexity and composition, the key protein involved in the transport of iron (as iron, heme or iron-siderophore) across the outer membrane is a TonB-dependent receptor with an overall structure presumably similar to that determined recently for *E. coli* FhuA or FepA. The receptors are potentially ideal vaccine targets in view of their critical role in survival in the host.

Helicobacter pylori is known to be an etiologic agent of gastritis and peptic ulcer disease in humans. Human LF supported full growth of the *H. pylori* in media lacking other iron sources, but neither human TF, bovine LF, nor hen ovoTF served as a source for iron (Husson et al., 1993). Since hLF is found in significant amounts in human stomach resections with superficial or atrophic gastritis, the iron acquisition system of *H. pylori* by the hLF receptor system may play a major role in the virulence of *H. pylori* infection. Most *H. pylori* strains also seem to produce extracellular siderophores (Illingworth et al., 1993).

The ability of malleobactin to mobilize iron from LF and TF was examined in an equilibrium dialysis assay in the absence of bacteria (Yang et al., 1993). Malleobactin was capable of removing iron from both LF and TF at pH values of 7.4, 6.0, and 5.0. However, the levels of iron mobilization were greater for TF than for LF at all the pH values used in the assay. *Bordetella bronchiseptica* uses a hydroxamate siderophore for removal of iron from LF and TF rather than relying upon a receptor for these host iron-binding proteins (Foster & Dyer, 1993).

Moraxella (Branhamella) catarrhalis, a mucosal pathogen closely related to *Neisseria* species, is a prominent cause of otitis media in young children and lower respiratory tract infections in adults. Campagnari et al. (1994) demonstrated that *M. catarrhalis* obtains iron from LF and TF and also maintains growth with ferric nitrate *in vitro*. Furthermore, when *M. catarrhalis* is grown under iron-limited conditions, the bacteria express new outer membrane proteins that are not detected in membranes of organisms cultured in an iron-rich environment. These iron-repressible proteins may be important for the acquisition and utilization of iron *in vivo*, which could allow *M. catarrhalis* to colonize and survive on human mucosal surfaces.

Vulnibactin, a siderophore produced by *Vibrio vulnificus*, has been shown to sequester TF- or LF-bound iron for growth (Okujo et al., 1996). Comparative studies with the strain producing vulnibactin and its exocellular protease-deficient mutant revealed the involvement of the protease in addition to vulnibactin could be effective in the utilization of Fe(III) bound to TF and LF. It appears that the protease causes cleavage of these proteins, thereby making bound iron more accessible to vulnibactin. In response to environmental iron stress, *Vibrio cholerae* produces the siderophore vibriobactin as well as a number of iron-induced outer membrane proteins (Tashima et al., 1996).

Leishmania chagasi, the cause of South American visceral leishmaniasis, requires iron for its growth. Wilson et al. (1994) reported the ability of promastigote forms of *L. chagasi* to take up ^{59}Fe chelated to either TF or LF, although uptake from ^{59}Fe-LF occurred more rapidly. ^{59}Fe uptake from either ^{59}Fe-TF or ^{59}Fe-LF was inhibited by a 10-fold excess of unlabeled holo-LF, holo-TF, apo-LF, apo-TF, or iron nitrilotriacetate but not ferritin or bovine serum albumin. There was no evidence for a role for parasite-derived siderophores or proteolytic cleavage of holo-LF or holo-TF in the acquisition of iron by

promastigotes. This capacity to utilize several iron sources may contribute to the organism's ability to survive in the diverse environments it encounters in the insect and mammalian hosts.

Gardnerella vaginalis could acquire iron from hLF, but not from hTF (Jarosik et al., 1998). Siderophore production was detected in *G. vaginalis* strains and SDS-PAGE of the cytoplasmic membrane proteins isolated from *G. vaginalis* grown under iron-replete and iron-restricted conditions revealed several iron-regulated proteins ranging in molecular mass from 33- to 94-kDa.

VIII. MICROBIAL INTERACTIONS

LF could bind to microbial surface via an array of specific and non-specific interactions. Interactions of LF with specific microbial target sites could lead to events either for promoting host defense (microbial elimination) or microbial virulence (iron acquisition by pathogen) (Naidu & Arnold, 1997). This chapter shall focus the role of LF-binding microbial targets in antimicrobial susceptibility. Various LF-binding microbial targets have been listed in *TABLE 5*.

A. LF-binding targets

Specific LF-binding targets were identified on a variety of bacterial pathogens. The binding components in most microorganisms were proteins, and a few lectin-type interactions were also reported.

1. Staphylococcus spp. Bovine LF could bind to the following staphylococcal species associated with bovine intramammary infections: *S. epidermidis, S. warneri, S. hominis, S. xylosus, S. hyicus*, and *S. chromogenes* (Naidu et al., 1990). The bLF-binding mechanism was specific, with affinity constants (K_a values) ranging between 0.96 mM and 11.9 mM. The numbers of bLF-binding sites per cell, as determined by using Scatchard analysis, were as follows: *S. epidermidis*, 3,600; *S. warneri*, 1,900; *S. hominis*, 4,100; *S. xylosus*, 4,400; *S. hyicus*, 6,100; and *S. chromogenes*, 4,700. The bLF- binding receptors of the six coagulase-negative staphylococcal species demonstrated marked differences in patterns of susceptibility to proteolytic or glycolytic enzyme digestion and to heat or periodate treatment. These data suggest that the bLF-binding components in *S. epidermidis* and *S. warneri* are proteins containing glycosidyl residues. In the remaining four species, the proteinaceous nature of the bLF-binding component was evident, but the involvement of glycosidyl residues was not clear.

Naidu and co-workers (1991) investigated the hLF binding property of 489 strains of *S. aureus* isolated from various clinical sources. The hLF binding was common among *S. aureus* strains associated with furunculosis (94.3%), toxic shock syndrome (94.3%), endocarditis (83.3%) and septicaemia (82.8%) and other (nasal, vaginal or ocular) infections (96.1%). Naidu et al. (1992) also characterized the hLF-staphylococcal interaction in *S. aureus* strain MAS-89. The binding of ^{125}I-hLF to strain MAS-89 reached saturation in less than 90 min and was maximal between pH 4 and 9. Unlabelled hLF displaced ^{125}I-HLF binding. Various plasma and subepithelial matrix proteins, such as IgG, fibrinogen, fibronectin, collagen and laminin, which are known to interact specifically with *S. aureus*, did not interfere with hLF binding. The Scatchard plot was non-linear; that

TABLE 5: Specific binding of LF to various microbial cell surfaces - density, mass, binding-affinity (association constant) of cellular target sites

Microorganism	LF type	LF-binding cellular target site	Sites/cell	Mass	Affinity (K_a)	Reference
Actinobacillus actinomycetemcomitans	bLF/hLF	Heat-modifiable OMPs	ND	29-, 16.5-kDa	880 /1,800 nM	Alugupalli et al.,1995
Aeromonas hydrophila CCUG14551	bLF/hLF	Outer membrane proteins / porins	ND	30-, 40-, 60-kDa	ND	Kishore et al., 1991
Bordetella bronchseptica	hLF	Cell surface protein	ND	32-kD	ND	Menozzi et al., 1991
Bordetella pertussis	hLF	Cell surface protein	ND	27-kD	ND	Menozzi et al., 1991
Clostridium sp.	bLF	Surface layer protein	ND	33-kDa	ND	Tomita et al., 1998
Escherichia coli E34663	bLf/hLF	Outer membrane proteins (OMPs)	5,400	ND	140 nM	Naidu et al., 1991
Escherichia coli E34663	bLf/hLF	Porins Omp-C, Omp-F & Pho-E		37-kDa		Erdei et al., 1994
Escherichia coli O55B5	bLf/hLF	Lipopolysaccharide	ND		4 / 390 nM	Elas-Rochard et al., 1995
Haemophilus influenza KC548	hLF	Membrane protein	ND	106-kD, 105-kD	ND	Schryver, 1989
Helicobacter pylori	bLf/hLF	Heat-shock protein	ND	60-kDa	2,880 nM	Amini et al., 1996
	hLF	Outer membrane protein	ND	70-kDa	ND	Dhaenens et al., 1997
Mycoplasma pneumoniae	hLF	Cell membrane	10,000	ND	20 nM	Tryon & Baseman, 1987
Nesseria gonorrhoeae	hLF	Outer membrane receptor	ND	103-kD	ND	Biswas & Sparling, 1995
Neisseria meningitidis	hLF	Outer membrane receptor	ND	105-kD	ND	Schryver & Morris, 1988
	hLF	OMP - *IroA* protein	ND		ND	Pettersson et al., 1994
Porphyromonas gingivalis	hLF	Fimbriae, lectin-type interaction	ND	ND	ND	Sojar et al., 1998
Prevotella intermedia 4H	bLf/hLF	Cell surface protein	45,000	62-kDa	550 nM	Kalfas et al., 1991; 1992
Pseudomonas aeruginosa	hLF	Outer membrane protein	ND	48-kD, 25-kD	ND	Carnoy et al., 1994
Salmonella abony NCTC6017	bLF/hLF	Outer membrane proteins / porins	ND	38-, 35-, 32-kDa	ND	Naidu & Arnold, 1994
Salmonella dublin NCTC9676	bLF/hLF	Outer membrane proteins / porins	ND	38-kDa	ND	Naidu & Arnold, 1994
Salmonella hartford HNCMB10063	bLF/hLF	Outer membrane proteins / porins	ND	37-, 35-, 25-kDa	ND	Naidu & Arnold, 1994
Salmonella panama NCTC5774	bLF/hLF	Outer membrane proteins / porins	ND	37-kDa	ND	Naidu & Arnold, 1994
Salmonella pullorum NCTC5776	bLF/hLF	Outer membrane proteins / porins	ND	38-, 35-kDa	ND	Naidu & Arnold, 1994
Salmonella rostock NCTC5767	bLF/hLF	Outer membrane proteins / porins	ND	39-kDa	ND	Naidu & Arnold, 1994
Salmonella kentucky NCTC5799	bLF/hLF	Outer membrane proteins / porins	ND	38-, 35-, 33-kDa	ND	Naidu & Arnold, 1994
Salmonella thompson NCTC5740	bLF/hLF	Outer membrane proteins / porins	ND	38-, 35-, 33-kDa	ND	Naidu & Arnold, 1994
Salmonella typhimurium ATCC13311	bLF/hLF	Outer membrane proteins / porins	ND	38-kDa	ND	Naidu & Arnold, 1994
Salmonella virchow NCTC 5742	bLF/hLF	Outer membrane proteins / porins	ND	38-, 35-, 33-kDa	ND	Naidu & Arnold, 1994
Shigella flexneri M90T	hLF	OMPs, porins	4,800	39-, 22-, 16-kDa	690 nM	Tigyi et al., 1992
Shigella flexneri M90T	bLF	OMPs, porins	5,700	39-, 22-, 16-kDa	104 nM	Tigyi et al., 1992
Staphylococcus aureus MAS89	hLF	Proteoglycan component	5,700	67-kDa / 62-kDa	27 nM	Naidu et al., 1992
Staphylococcus chromogenes AD1	bLF	Cell wall protein	4,700	ND	2,500 nM	Naidu et al., 1990
Staphylococcus epidermidis AF9	bLF	Peptidoglycan component	3,600	ND	11,900 nM	Naidu et al., 1990
Staphylococcus hominis AF93	bLF	Cell wall protein	4,100	ND	3,800 nM	Naidu et al., 1990

Table 5 (Cont.): Specific binding of LF to various microbial cell surfaces - density, mass, binding-affinity (association constant) of cellular target sites

Microorganism	LF type	LF-binding cellular target site	Sites/cell	Mass	Affinity (K_a)	Reference
Staphylococcus hyicus AC166	bLF	Cell wall protein	6,100	ND	960 nM	Naidu et al., 1990
Staphylococcus warneri AF101	bLF	Peptidoglycan component	1,900	ND	3,100 nM	Naidu et al., 1990
Staphylococcus xylosus AG12	bLF	Cell wall protein	4,400	ND	3,300 nM	Naidu et al., 1990
Treponema denticola	hLF	Cell surface proteins	ND	50-, 35-kDa	ND	Staggs et al., 1994
Treponema pallidum	hLF	Cell surface proteins	ND	49-, 34-, 29-kDa	ND	Staggs et al., 1994
Tritrichomonas foetus	hLF	Cell wall receptor	170,000	ND	3,600 nM	Tachezy et al., 1996
Trichomonas vaginalis	hLF	Cell surface receptor	90,000	ND	1,000 nM	Peterson & Alderete, 1984
Trypanosoma cruzi	hLF	Amastigote cell wall	1,000,000	ND	ND	Lima & Kierszenbaum, 1985

implied a low affinity (K_a:155 nM) and a high affinity (K_a:270 nM) binding mechanism. About 5,700 hLF-binding sites/cell were estimated. The staphylococcal hLF- binding protein (hLF-BP) was partially susceptible to proteolytic enzymes or periodate treatment and was resistant to glycosidases. An active hLF-BP with an apparent M_r of about 450-kDa was isolated from strain MAS-89 cell lysate by ion- exchange chromatography on Q-sepharose. In SDS-PAGE, the reduced hLF-BP was resolved into two components of 67- and 62-kDa. The two components demonstrated a positive reaction with hLF-HRPO in a Western blot. These data established a specific receptor for hLF in *S. aureus*.

2. Streptococcus spp. The exposure of *Streptococcus agalactiae* to bLF resulted in the binding of this protein to all the 12 strains of bovine origin tested, and also, although to a lesser degree, to the five tested strains of human origin (Rainard, 1992). The binding of LF was slightly affected by cultivation conditions, and appeared to be heat-stable. The binding of biotinylated LF was inhibited by unlabeled-LF but not by BSA.

Hammerschmidt et al. (1999) demonstrated specific binding of hLF to *Streptococcus pneumoniae*. Pretreatment of pneumococci with proteases reduced hLF binding markedly, indicating that the hLf receptor is proteinaceous. Binding assays performed with 63 clinical isolates belonging to different serotypes showed that 88% of the tested isolates interacted with hLF. Scatchard analysis showed the existence of two hLF-binding proteins with dissociation constants of 57 nM and 274 nM. The receptors were purified by affinity chromatography, and internal sequence analysis revealed that one of the *S. pneumoniae* proteins was homologous to pneumococcal surface protein A (PspA). The function of PspA as an hLF-binding protein was confirmed by the ability of purified PspA to bind hLF and to competitively inhibit hLF binding to pneumococci. *S. pneumoniae* may use the hLF-PspA interaction to overcome the iron limitation at mucosal surfaces, and this might represent a potential virulence mechanism.

3. Vibrio spp. Binding of LF to non-invasive *Vibrio cholerae* was reported (Ascencio et al., 1992). Iron-binding glycoproteins such as ferritin, TF, haemoglobin, and myoglobin moderately inhibited the LF interaction with the vibrios. Monosaccharides (N-acetyl glucosamine, mannose, galactose, and fucose), and other glycoproteins such as fetuin and orosomucoid also moderately inhibited the binding. *V. cholerae* showed a cell surface associated proteolytic activity which cleaved the cell-bound ^{125}I-labeled LF.

4. Helicobacter pylori: The interactions of *Helicobacter pylori* spiral and coccoid forms with LF was reported by Khin et al. (1996). The coccoid forms of 14 strains of *H. pylori* showed significant hLF binding (median 26%), found to be specific and was inhibited by unlabeled hLF and bLF.

Amini et al. (1996) reported the binding of bLF to a 60-kDa heat shock protein of *H. pylori*. Binding ability was related to human immunoglobulin G because bLF binding proteins were isolated by extraction of cell surface associated proteins with distilled water, applied on IgG-Sepharose and nickel sulfate chelate affinity chromatography. Binding was demonstrated by Western blot after purified protein was digested with alpha-chymotrypsin and incubated with peroxidase-labeled bLF. Binding was inhibited by unlabeled bLF, lactose, rhamnose, galactose, and two iron-containing proteins, ferritin and haptoglobin. Carbohydrate moieties of bLF seem to be involved in binding because glycoproteins with similar carbohydrate structures strongly inhibited binding. Scatchard plot

analysis indicated a binding affinity (K_a) of 2.88 μM. In addition, binding of *H. pylori* cells to bLF was enhanced when bacteria treated with pepsin or α-chymotrypsin after isolation from iron-restricted and iron-containing media.

5. Oral bacteria. Interaction of LF with *Actinobacillus actinomycetemcomitans* was reported (Alugupalli et al., 1995). The binding of hLF and bLF reached maximum within 1 h. LF binding to the bacterium was pH-dependent and reversible. Scatchard analysis indicated the existence of two different types of binding sites on the bacterium, one with a high affinity constant (K_a:0.880 μM) and the other with a low affinity (K_a:1.8 μM). Bacteria in the exponential phase of growth showed higher binding than cells in the stationary phase. Bacteria grown in medium containing serum and/or lysed erythrocytes bound LF to a lesser extent. Heat-inactivated serum, lysed erythrocytes and other proteins such as mucin and laminin inhibited LF binding to *A. actinomycetemcomitans* in a competitive binding assay. SDS-PAGE and Western blot analysis revealed LF-reactive protein bands at 29-kDa and 16.5-kDa in the CE and OM of *A. actinomycetemcomitans*. The 29-kDa band displayed a heat-modifiable LF-reactive form with a molecular weight of 34-kDa. Neither proteinase K-treated cell envelope nor LPS of this bacterium showed reactivity with LF. These data suggest a specific LF interaction with OMPs of *A. actinomycetemcomitans*.

A LF-binding protein with an estimated molecular mass of 57-kDa was identified in the cell envelope of *Prevotella intermedia* by SDS-PAGE and Western- blot analysis (Alugupalli et al., 1994). Peroxidase-labeled bLF and hLF showed similar specific binding to this protein.

B. Porins

Porins are a well conserved heat-modifiable, pore-forming outer membrane proteins (OMPs) of the family *Enterobacteriaceae*. Porins are also reported in certain Gram-negative bacteria. Porins exist as trimers in the outer membrane usually surrounded by nine molecules of LPS. Porins are suggested to play an essential role in the transport of various solutes across the Gram-negative OM. Porins also serve as receptors for certain bacteriophages and colicins (see reviews: Lugtenberg & van Alpen, 1983; Nikaido & Vaara, 1985; Nikaido, 1989). Naidu and co-workers have reported the role of porins as LF anchorages in *E.coli* and other bacterial members of the family *Enterobacteriaceae* (Kishore et al., 1991; Tigyi et al., 1992; Erdei et al., 1993; Naidu & Arnold, 1994).

1. Escherichia coli. The degrees of hLF and bLF binding in 169 *E. coli* strains isolated from human intestinal infections, and in an additional 68 strains isolated from healthy individuals, were examined in a ^{125}I-labelled protein binding assay (Naidu et al., 1991). The binding was expressed as a percentage calculated from the total labelled ligand added to bacteria. The hLF and bLF binding to *E. coli* was in the range 3.7 to 73.4% and 4.8 to 61.6%, respectively. Enterotoxigenic (ETEC) strains demonstrated a significantly higher hLF binding than enteropathogenic (EPEC), enteroinvasive (EIEC), enterohaemorrhagic (EHEC) strains or normal intestinal *E. coli* isolates. EPEC strains belonging to serotypes O44 and O127 demonstrated significantly higher hLF binding compared to O26, O55, O111, O119 and O126 serotypes. No significant differences in the degree of hLF or bLF binding were found between aerobactin-producing and non-producing strains. The interaction was further characterized in a high LF-binding EPEC strain, E34663 (serotype O127). The binding was stable in the pH range 4.0 to 7.5, did not dissociate in

the presence of 2M NaCl or 2M urea, and reached saturation within 2-h. Unlabelled hLF and bLF displaced the ^{125}I-hLF binding to E34663 in a dose-dependent manner. Apo- and iron- saturated forms of LF demonstrated similar binding to E34663. Among various unlabelled subepithelial matrix proteins and carbohydrates tested (in 4-log excess) only fibronectin and fibrinogen caused a moderate inhibition of ^{125}I-hLF binding. According to Scatchard plot analysis, 5,400 hLF-binding sites/cell, with an affinity constant (K_a) of 140 nM, were estimated in strain E34663. These data establish the presence of a specific LF-binding mechanism in *E. coli*.

Gado et al. (1991) reported that *E. coli* with low hLF binding are insusceptible to group A (A, E1, E2, E3, E6, and K) and group B (B, D, Ia, Ib, and V) colicins. Conversely, a spontaneous hLF high-binding variant demonstrated an increased susceptibility to both colicin groups. Colicin-insusceptible *E. coli* wild-type strains 75ColT, 84ColT, and 981ColT showed a low degree of hLF binding, i.e., 4, 8, and 10%, respectively. The hLF binding capacity was high in the corresponding colicin-susceptible mutants 75ColS (43%), 84ColS (32%), and 981ColS (43%). Furthermore, hLF low- (< 5%) and high- (> 35%) binding *E. coli* clinical isolates (10 in each category) were tested for susceptibility against 11 colicins. Colicin V susceptibility did not correlate with hLF binding in either categories. However, with the remaining colicins, three distinct hLF-binding, colicin susceptibility patterns were observed; (i) 10 of 10 hLF low-binding strains were colicin insusceptible, (ii) 6 of 10 hLF high-binding strains were also colicin insusceptible, and (iii) the remaining hLF high binders were highly colicin susceptible. Certain proteins in the cell envelope and outer membrane of wild-type H10407 (hLF low binder, colicin insusceptible) showed a lower mobility in SDS-PAGE compared to the corresponding proteins of mutant H10407(LF) (hLF high binder, colicin susceptible). These mobility differences were also associated with hLF-binding proteins in Western blot (ligand blot) analysis. The wild type showed a smooth form of LPS with a distinct ladder of O-chains, compared to the rough LPS of the mutant.

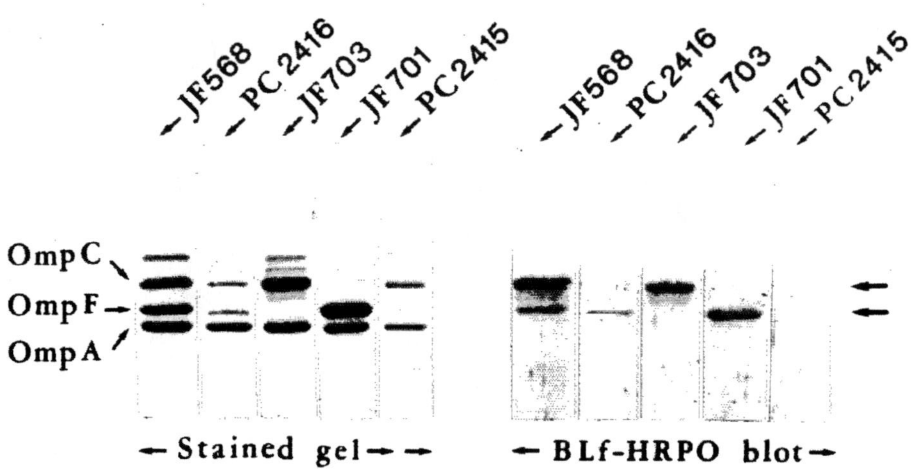

FIGURE 8. LF interaction with porins of *E.coli*. Outer membrane analyses of wild-type (JF568) and porin-deficient mutants (PC2416 express *PhoE* only; JF703 express *OmpC* only; JF701 express *OmpF* only; and PC2415 express none of the three porins) of *E.coli* by urea-SDS-PAGE and western blotting with HRPO-labeled bLF [from Erdei et al. (1994) with permission from the American Society for Microbiology].

Certain strains of *E. coli* (bacterial whole cells) demonstrate specific interaction with ^{125}I-LF. A band with a mass of approximately 37-kDa, which was reactive with horseradish peroxidase (HRPO)-labeled LF, was identified in the boiled cell envelope and outer membrane preparations of an LF-binding *E. coli* strain, E34663, and a non-LF-binding strain, HH45, by SDS-PAGE and Western blotting (Erdei et al., 1994). Such a band was not detected in the unboiled native cell envelope and outer membrane preparations. The molecular mass and the property of heat modifiability suggested that the LF-binding proteins were porins. The native trimeric form of porin OmpF isolated from strain B6 and its dissociated monomeric form both reacted with HRPO-labeled LF and with monoclonal antibodies specific for OmpF. Furthermore, by using *E. coli* constructs with defined porin phenotypes, OmpF and OmpC were identified as the LF-binding proteins by urea-SDS-PAGE and Western blotting and by ^{125}I-LF binding studies with intact bacteria (*FIGURE 8*). These data established that LF binds to porins, a class of well-conserved molecules common in *E. coli* and many other Gram-negative bacteria. However, in certain strains of *E. coli* these pore-forming proteins are shielded from LF interaction.

2. Salmonella spp. Interaction of LF with the CE and OM of *Salmonella typhimurium*-type strain ATCC13311 was tested by SDS-PAGE and Western-blot analyses (Naidu & Arnold, 1994). The HRPO-labeled bLF and hLF both recognized a heat-modifiable protein with an estimated molecular mass of 38-kDa in the OM. Simultaneous immunoblotting with an antiporin monoclonal antibody specific for a conserved porin domain in members of *Enterobacteriaceae* confirmed that the LF-binding protein is a porin. Such LF-binding porin proteins (37- to 39-kDa range) were readily detected in nine other common *Salmonella* species: *S. dublin, S. panama, S. rostock, S. abony, S. hartford, S, kentucky, S. pullorum, S. thompson,* and *S. virchow*. The latter six species also demonstrated one to three weak LF-reactive bands of low molecular weight in their CE.

3. Shigella flexneri. Tigyi et al. (1992) reported the interaction of LF with dysentry pathogen, *Shigella flexneri*. The interaction was specific, and approximately 4,800 hLF binding sites (K_a:690 nM) or approximately 5,700 bLF binding sites (K_a:104 nM) per cell were estimated in strain M90T by a Scatchard plot analysis. The native CE and OM did not reveal LF-binding components in SDS-PAGE. However, after being boiled, the CE and OM preparations showed three distinct horseradish peroxidase-LF reactive bands of about 39-, 22-, and 16-kDa. The 39-kDa component was also reactive to a monoclonal antibody specific for porin (PoI) proteins of members of the family *Enterobacteriaceae*. The LF binding protein pattern was similar with bLF or hLF, for Crb+ and Crb- strains. The protein-LF complex was dissociable by KSCN or urea and was stable after treatment with NaCl. Variation (loss) in the O chain of LPS markedly enhanced the LF-binding capacity in the isogenic rough strain SFL1070-15 compared with its smooth parent strain, SFL1070. These data establish that LF binds to specific components in the bacterial OM; the heat-modifiable, anti-PoI-reactive, and LPS-associated properties suggested that the LF-binding proteins are porins in *S. flexneri*.

4. Aeromonas hydrophila. The interaction of LF with *Aeromonas hydrophila* was tested in a ^{125}I-labeled protein- binding assay (Kishore et al., 1991). The LF binding was characterized in type strain *A. hydrophila* ssp. *hydrophila* CCUG 14551. The hLF and

bLF binding reached a complete saturation within 2 h. Unlabeled hLF and bLF displaced ^{125}I- hLF binding in a dose-dependent manner, and more effectively by the heterologous (1 μg for 50% inhibition) than the homologous (10 μg for 50% inhibition) ligand. Apo- and holo-forms of hLF and bLF both inhibited more than 80%, while mucin caused approx. 50% inhibition of the hLF binding. Various other proteins (including TF) or carbohydrates did not block the binding. Two hLF-binding proteins with an estimated molecular masses of 40-kDa and 30-kDa were identified in a boiled-CE preparation, while the unboiled CE demonstrated a short-ladder pattern at the top of the separating gel and a second band at approx. 60-kDa position. These data establish a specific interaction of LF and the LF-binding proteins seem to be porins in *A. hydrophila*.

C. Lipopolysaccharides (LPS)

1. Interactions. LF was reported to bind lipid A and intact LPS of *E. coli, Klebsiella pneumoniae, Pseudomonas aeruginosa, Neisseria meningitides* and *Haemophilus influenzae* (Appelmelk et al. 1994). LF binding to LPS was inhibitable by lipid A and polymyxin B but not by KDO (3-deoxy-D-manno-octulosonate), glycoside residues present in the inner core of LPS. Binding of LF to lipid A was saturable, and with an affinity constant of 2 nM.

Elass-Rochard et al (1995) reported the presence of two *E. coli 055B5* LPS-binding sites on hLF: a high-affinity binding site (K_a:3.6 nM) and a low-affinity binding site (K_a:390 nM). Bovine LF, which shares about 70% amino acid sequence identity with hLF, showed similar interaction with LPS. Human serum TF, which is known to bind LPS, caused only 12% inhibition of hLF-LPS interaction, suggesting different binding domains on LPS. Binding and competitive binding experiments performed with the N-tryptic fragment (residues 4-283), the C-tryptic fragment (residues 284-692) and the N2-glycopeptide (residues 91-255) isolated from hLF have demonstrated that the high-affinity binding site is located in the N-terminal domain I of hLF, and the low-affinity binding site is present in the C-terminal lobe. The inhibition of hLF-LPS interaction by a synthetic octadecapeptide corresponding to residues 20-37 of hLF and bLFcin (residues 17-41), a proteolytic fragment from bLF, revealed the importance of the 28-34 loop region of hLF and the homologous region of bLF for LPS binding. Direct evidence that this amino acid sequence is involved in the high-affinity binding to LPS was demonstrated by assays carried out with EGS-loop hLF, a recombinant hLF mutated at residues 28-34.

After differentiation, HL-60 cells showed a twofold increase of LF- binding sites with no difference in the specificity or affinity of LF between pre- and post-differentiated cells (Miyazawa et al, 1991). CD11a, CD11b, and CD11c Ag, which have been associated with specific binding sites for LPS on monocytes/macrophages, were also increased 3- to 4-fold after differentiation. With the use of this system, the effect of LPS on LF binding was tested. At 37°C, LPS enhanced LF binding on HL-60 cells, especially after differentiation. Conversely, at 4°C, LPS inhibited LF binding. There was little effect of temperature on LF binding in the absence of LPS. In the presence of polymyxin B sulfate, the enhanced LF binding by LPS was abrogated. In addition, pretreatment with mAbCD11 and/or mAb5D3, which are associated with or directed against candidate LPS receptors reduced LF binding. Cross-linking studies using an iodinated, photoactivatable LPS derivative ([^{125}I] ASD-LPS) demonstrated specific binding of LPS to LF. These data indicated a dichotomous nature of LF binding on monocyte/macrophage-differentiated HL-60 cells;

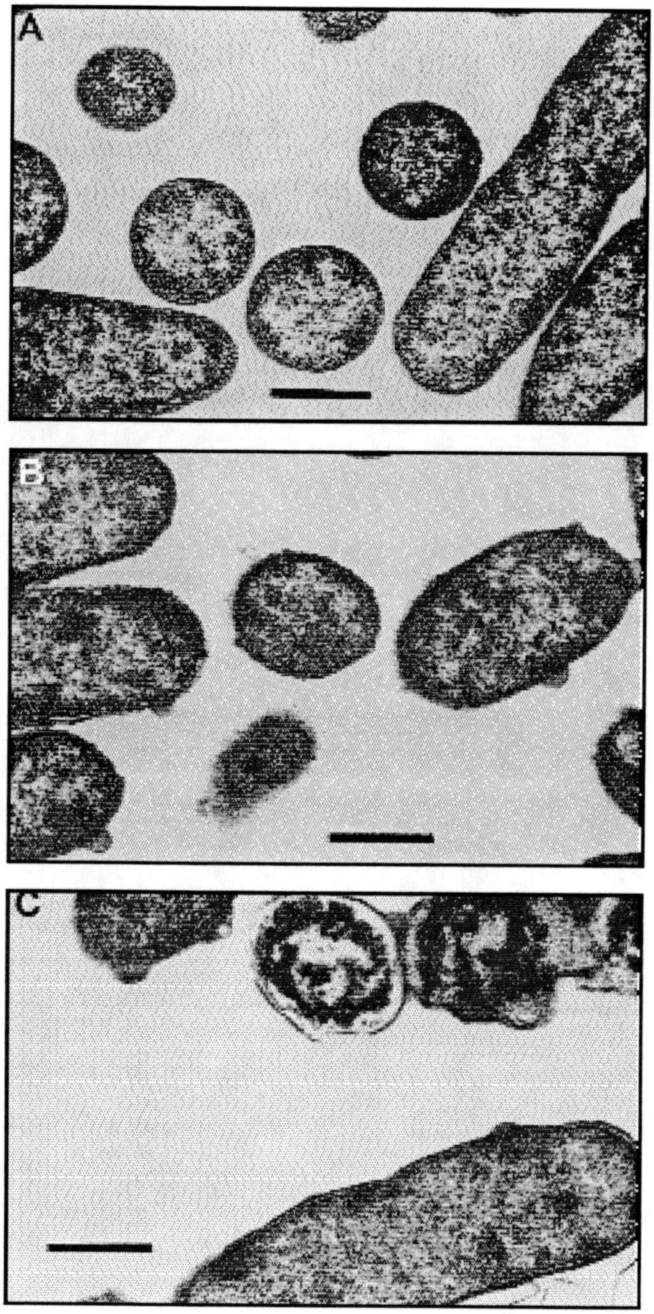

FIGURE 9. LFcin effects on *E.coli* CL99 I-2 cells (TEM: Bars, 50 nm). Bacteria were incubated for 2-h in 1% Bacto-peptone broth (A); or broth containing 0.1 mg/ml of bLFcin for 0-h (B) or 2-h (C). Bacteria exposed to bLFcin show altered cell membrane morphology with the appearance of membrane blisters. After 2-h incubation a large amount of cell debris was present and a number of remaining cells appear to have a clumping or coagulation of cytoplasmic elements in addition to membrane blistering [from Yamauchi et al., (1993) with permission from the American Society for Microbiology].

one being mediated by specific LF receptors whereas the other is apparently mainly via LPS receptors after formation of an LF-LPS complex.

LF could inhibit iron-catalyzed formation of hydroxyl radical in the presence of LPS at pH 7.4 and 4.5 (Cohen et al, 1992). Low concentrations of LPS could prime neutrophils toward enhanced function, such as formation of stimulated superoxide anion. LF inhibited LPS priming of neutrophils if LPS contamination of the protein (provided by commercial suppliers) was first reduced. Inhibition of LPS priming was observed whether apo-LF or Fe-LF was used. Similar inhibition of LPS priming was observed when neutrophils were incubated with other serum proteins (e.g., albumin, apoTF, or iron-saturated TF). These results indicated that LPS should not be expected to affect the free radical biology of LF, which is a crucial physiologic function of this protein.

PMN-induced LF could inactivate LPS, thereby blocking the ability of LPS to prime fresh PMNs for enhanced fMLP-triggered release of superoxide (Wang et al, 1995). Neutrophils (10^6 cells/ml) inactivation of LPS (10 ng/ml) took 30 min in a kinetic fashion. In addition, LF isolated from PMN population also required 30 min to inactivate LPS, indicating inherently slow binding of LF to LPS. Mononuclear cells failed to inactivate LPS under the same conditions. Studies with isotope-labeled LPS showed that inactivated LPS remained in the medium and was not taken up or destroyed by the PMNs during inactivation. Bovine LF was also reported to diminish the inflammatory reactions induced by *Mycobacterium bovis* (BCG) in a mouse model (Zimecki & Machnicki, 1994). Human LF, bLF and LFcin-B were found to suppress the IL-6 response in a monocytic cell line (THP-1) when stimulated by LPS (Mattsby-Baltzer et al, 1996). The suppression of bLF was similar to or higher than that of hLF. LFcin-B was the strongest inhibitor of the LPS-induced IL-6 response. For hLF, the strongest inhibition was observed when added 15-30 min after the addition of LPS. Addition of LF before the LPS induced an approximately 45% reduction of the IL-6 response. The results suggest an anti-inflammatory activity of hLF, bLF, and bLFcin through their suppressive effects on the cytokine release.

2. LF-induced LPS release - effects on microbial OM permeability. Ellison and co-workers (1988) reported that the iron-binding proteins could damage the gram-negative outer membrane and alter bacterial outer membrane permeability in a manner similar to that of the chelator EDTA. Studies in barbital-acetate buffer showed that EDTA and hLF cause significant release of radio-labeled LPS from a UDP-galactose epimerase-deficient *E. coli* mutant and from wild-type *S. typhimurium* strains. The LPS release was blocked by iron saturation of hLF, occurred between pH 6 and 7.5, was comparable for bacterial concentrations from 10^4 to 10^7 CFU/ml, and increased with increasing hLF concentrations. Studies using Hanks balanced salt solution lacking calcium and magnesium showed that TF could cause LPS release. Additionally, both hLF and TF increased the antibacterial effect of a sub-inhibitory concentration of rifampin, a drug excluded by the bacterial outer membrane.

Bovine LF and LFcin also reported to release intrinsically labeled [^{3}H] LPS from three bacterial strains, *E. coli CL99 1-2*, *S. typhimurium SL696*, and *Salmonella montevideo SL5222* (Yamauchi et al, 1993). Under most conditions, more LPS are released by LFcin, the peptide fragment than by whole bLF. In the presence of either, LF or LFcin there is increased killing of *E. coli CL99 1-2* by lysozyme. Bovine LF and LFcin have the ability to bind to free intrinsically labeled [^{3}H] LPS molecules, similar to hLF. In addition to these effects, whereas bLF was at most bacteriostatic, LFcin demonstrated consistent

bactericidal activity against gram-negative bacteria. This bactericidal effect is modulated by the cations Ca^{2+}, Mg^{2+}, and Fe^{3+} but is independent of the osmolarity of the medium. Transmission electron microscopy of bacterial cells exposed to LFcin show the immediate development of electron-dense membrane blisters (*FIGURE 9*).

LF shows synergism with lysozyme; each protein alone is bacteriostatic, however, the mixture elicit a bactericidal effect for strains of *E. coli, S. typhimurium*, and *Vibrio cholerae* (Ellison & Giehl, 1991). The cidal activity is dose dependent, blocked by iron saturation of LF, and inhibited by high calcium levels, although LF does not chelate calcium. Growth media conditions inhibit completely or partially the effect of LF and lysozyme; the degree of inhibition correlate with media osmolarity. Dialysis chamber studies indicated that bacterial killing required direct contact with LF, and experiments with purified LPS suggested that this related to direct LPS- binding by the protein.

3. LPS induction of LF production in vivo. When incubated with *Salmonella typhimurium* LPS at 37°C, human PMN suspended in serum-free buffer releases the specific granule constituent LF into the surrounding medium (Koivuranta-Vaara et al., 1987). Release of LF from PMN vary with the concentration of LPS as well as with the duration of incubation and is not accompanied by significant release of the cytoplasmic enzyme lactate dehydrogenase. LPS-induced release of LF from PMN was augmented significantly when cell suspensions were supplemented with additional monocytes and lymphocytes. Only monocytes, however, secreted significant amounts of LF-releasing activity (in a time- and concentration- dependent manner) when incubated separately with LPS. LF-releasing activity was heat (80°C for 15 min) labile, eluted after chromatography on Sephadex G-100 with an apparent molecular weight of approximately 60-kDa, and was inhibited by antibodies to TNF-α. Thus, LPS-induced noncytotoxic release of LF from human PMN suspended in serum-free buffer is mediated, at least in part, by TNF-α derived from contaminating monocytes.

Injection of *Salmonella typhimurium* or LPS into mice resulted in a dose-dependent increase in plasma LF. Endotoxin challenge of normal and neutropenic mice showed a direct correlation of plasma LF level with the granulocyte count in peripheral blood. Physiological neutropenia did not inhibit the LF release (Sawatzki & Rich, 1989). *In vivo* studies by Gutteberg et al. (1989) measured the total serum iron, plasma LF and circulating leukocytes in piglets during the early phase of severe gram-negative septicemia and endotoxemia following infusion of LPS or *E. coli*. Iron dropped significantly during the first 30 min of LPS infusion from a median of 32 mM to 13.4 mM. A similar decrease in serum iron was observed at 120 min after the *E. coli* infusion. Plasma levels of LF increased significantly 120 min after the LPS infusion (6 mg/L compared to preinfusion value of 0.25 mg/L). After intravenous infusion of *E. coli* a significant rise of plasma LF was demonstrated in 30 min after bacterial infusion (2.1 mg/L compared to preseptic value of 0.8 mg/L). This increase was accompanied with a significant drop of circulating leukocytes (7.3 x 10^9/L compared to preinfusion, 17 x 10^9/L) in the piglets receiving *E. coli* by intravenous route. However, intraperitoneal inoculaltion of *E. coli* did not show any significant change of plasma LF. The rapid onset of hyposideremia during endotoxemia and *E. coli* septicemia appeared to correlate with the release of LF from granulocytes and the clearance of iron-bound LF from blood or peritoneal cavity. LPS-stimulation of heparinized human blood resulted in a significant rise of LF, TNF-α, and thromboplastin

levels (Gutteberg et al, 1990). LPS induced the secretion of LF from granulocytes; the levels were 2.1 mg/L (in 5 min) and 5.3 mg/L (in 60 min) compared to the control values of 0.2 mg/L and 1.3 mg/L, respectively. Systemic administration of bLF into mice 24 hours prior to intravenous challenge of LPS (50 mg), significantly lowered the serum concentration of TNF-α and IL-6 (Machnicki et al, 1993). Doses of bLF (<0.1 mg) and pretreatment of mice with bLF (days 6-2 or 12-2 hours) before LPS challenges were not effective.

Lipopolysaccharides (1 to 100 ng/ml) from *E. coli*, *S. typhimurium*, and *Serratia marcescens*, strongly enhanced growth inhibition of *Candida albicans* by human PMNs *in vitro*. Flow cytometry analysis demonstrated that LPS markedly augmented phagocytosis of *Candida* cells by increasing the number of yeasts ingested per neutrophil as well as the number of neutrophils capable of ingesting fungal cells. LPS activation caused augmented release of LF, and antibodies against LF effectively and specifically reduced the anti-*C. albicans* activity of both LPS-stimulated and unstimulated PMN. Northern (RNA blot) analysis showed enhanced production of mRNAs for IL-1b, TNF-α, and IL-6 and in neutrophils within 1 h of stimulation with LPS. The cytokines were also detected in the supernatant of the activated PMN, and their synthesis was prevented by pretreatment of LPS-stimulated PMN with protein synthesis inhibitors, such as emetine and cycloheximide. These inhibitors, however, did not block either LF release or the anti-*Candida* activity of LPS-stimulated PMN. These results demonstrate the ability of various bacterial LPS to augment neutrophil function against *C. albicans* and may contribute to the antifungal effect of PMN. Moreover, the ability to produce cytokines upon stimulation by ubiquitous microbial products such as the endotoxins points to an extraphagocytic, immunomodulatory role of PMN during infection.

D. DNA

LF was reported to interact with deoxyribonucleic acid (DNA) in both its double-stranded and single-stranded configurations (Bennett & Davis, 1982). Additional evidence for a preferential reactivity with dsDNA was provided by the enzymatic treatment of preformed dsDNA-LF and ssDNA-LF complexes with S1 endonuclease, and DNAse 1. DNase digestion alone liberated free LF. The interaction of LF with DNA partially inhibited the binding of anti-DNA antibodies from patients with SLE in a standard Farr assay. Furthermore, DNA-anti-DNA (labeled with ^{125}I-IgG) complexes could be dispersed *in vitro* by the addition of LF. It is hypothesized that the release of LF by neutrophils chemotactically attracted to DNA-anti-DNA complexes may act as a feedback loop to modulate the inflammatory response in SLE.

Studies indicated that the cell membrane DNA might have a novel role as a receptor for LF. Bennett et al (1983) demonstrated the binding of human ^{125}I-labeled LF to a population of adherent mononuclear cells (ADMC) and nonrosetting (E-) lymphocytes, which was abolished by prior treatment of the cells with deoxyribonuclease (DNase), but not ribonuclease (RNase). Saturable binding of LF at 0 °C was demonstrated for both E- and ADMC, with equilibrium dissociated constants of 0.76 x 10^{-6} M and 1.8 x 10^{-6} M, respectively. E- cells bound 2.5 x 10^7 and ADMC bound 3.3 x 10^7 molecules of LF at saturation. Cell membranes isolated from ADMC and E- reacted with ^{125}I-labeled LF and prior treatment of the membranes with DNase abolished the binding. This study provided evidence that cell membrane DNA acts as a binding site for exogenous LF.

Further investigations showed that binding of LF to the surface of human neutrophils was also dependent upon the presence of cell surface DNA (Bennett et al, 1986). This evidence of a DNA-LF interaction was obtained by the co-isolation of LF with DNA by both gel chromatography and affinity chromatography using Heparin Sepharose CL 6B. The interaction of LF with neutrophils was a saturable phenomenon with a K_a of 6.2 x 10^{-6} M and a maximum binding of 9.2 x 10^6 molecules per cell.

Various groups have consistently observed that LF interacts avidly with nucleic acids (Bennett & Davis, 1982; Bennett et al., 1986; Hutchens et al., 1989). To determine which region of the molecule is important in these interactions, solid-phase ligand binding assays were performed with LF from human milk (natural hLF) and N-terminally deleted hLF variants (van Berkel et al., 1997). Iron-saturated and natural hLF bound equally well to heparin, lipid A, human lysozyme and DNA. Natural hLF lacking the first two N-terminal amino acids (*Gly1-Arg2*) showed reactivities of one-half, two-thirds, one-third and one-third towards heparin, lipid A, human lysozyme and DNA respectively compared with N-terminally intact hLF. A lack of the first three residues (*Gly1-Arg2-Arg3*) decreased binding to the same ligands to one-eighth, one-quarter, one-twentieth and one-seventeenth respectively. No binding occurred with a mutant lacking the first five residues (*Gly1-Arg2-Arg3-Arg4-Arg5*). An anti-hLF monoclonal antibody (E11) that reacts to an N-lobe epitope including *Arg5* completely blocked hLF-ligand interaction. These results showed that the N-terminal stretch of four consecutive arginine residues, *Arg2-Arg3-Arg4-Arg5*, has a decisive role in the interaction of hLF with DNA, heparin, lipid A, and lysozyme.

Incubation of LF-Cu(II) complexes with supercoiled plasmid Bluescript II SK DNA rapidly form open circular or linear forms of DNA. LF with bound Cu(II) caused extensive degradation of yeast tRNA molecules in the presence of hydrogen peroxide. Covalent modification of surface-exposed histidyl residues by carboxyethylation with diethylpyrocarbonate abolished the LF-associated hydrolytic activity. These results indicated that LF-bound Cu(II) could facilitate the hydrolysis of DNA and RNA molecules. Copper-binding sites on LF possibly served as centers for repeated production of hydroxyl radicals via a Fenton-type Haber-Weiss reaction. Thus, it was speculated that an enhanced nuclease activity associated with elevated local concentrations of LF could potentiate microbial degradation (Zhao & Hutchens, 1994).

Recent studies showed that LF enters the cell from the serum and is transported into the nucleus where it binds DNA. Specific binding sites for LF were reported on K562 cells (estimate the number of binding sites and the dissociation constant). Western blotting analysis of K562 lysates identified a 120-kDa molecule as specific LF-binding receptor. This binding at the cell surface caused a temperature-dependant internalization of LF, which was immunologically detectable as a DNA-linked protein in nuclear extracts (Garre et al, 1992). Specific DNA sequences that could confer LF-induced gene transcription of a reporter gene have now been identified. RNA and DNA could inhibit LF enhancement of the activity of natural killer and lymphokine-activated killer cells *in vitro*. LF taken up by K562 human myelogenous leukemia cells appears in the nucleus where it is bound to DNA. This binding of LF to DNA occurs under stringent conditions with distinct sequence specificity, and that interaction between LF and these sequences intracellularly leads to transcriptional activation (He & Furmanski, 1995). This potential direct transcriptional function of LF is unique and remains to be confirmed in whole cells or tissues.

TABLE 6: Lactoferrin – Multifunctional activities

Cellular function	Reference
Inflammatory amplification and neutrophil aggregation	Oseas et al., 1981
Inhibition of antibody-mediated cytotoxicity	Nishiya and Horwitz, 1982
Specific growth stimulation of lymphocytes	Hashizume et al., 1983
Down regulation of myelopoiesis	Pelus et al., 1981; Broxmeyer et al., 1987
Complement activation by C3 convertase inhibition	Kulics & Kijlstra, 1987
Intestinal iron absorption	Cox et al., 1979; Davidson & Lönnerdal, 1988
Enterocyte proliferation and gut maturation	Nichols et al., 1990
Up-regulation of thymocyte maturation	Zimecki et al., 1991
Up-regulation of monocyte cytotoxicity	McCormick et al., 1991
Regulation of antibody production	Zimecki et al., 1991
Regulation of cytokine production	Crouch et al., 1992
Down-regulation of tumor necrosis factor	Machnicki et al., 1993
Prevention of hydroxy radical-mediated tissue injury	Britigan et al., 1994

IX. MULTIFUNCTIONAL ACTIVITIES *IN VIVO*

LF plays a critical regulatory role in various physiological pathways (*TABLE 6*). Though iron-binding is considered an important molecular property of LF, a number of cellular functions are independent of this metal binding property. Specific and non-specific interactions of LF with cells, co-existence with a variety of bio-molecules at different milieu, molecular heterogeneity and structural flexibility confers a spectrum of multifunctionality to the LF molecule *in vivo*.

A. Pathogen control *in vivo*

1. Bacteria. Zagulski et al. (1989) reported a protective effect of bLF when administered intravenously to mice 24 h before a challenge with a lethal dose of *E. coli*. About 70% of mice pretreated with bLF survived challenge. The survival rates in control mice treated with *E. coli* alone and pretreated with bovine serum albumin were 4 and 8%, respectively. Similar protection was also observed with hLF administration. Sufficient amounts of ferric ions were given to mice, in single and multiple doses, for full serum TF saturation 30 min before or after *E. coli* administration. The multiple dose of ferric ions did not change considerably the survival rate of mice pretreated with bLF. In contrast, a single dose of ferric ions gradually decreased the survival rate of the mice after the first week of experiment. From day-14 this decrease was statistically significant in all groups of mice treated with a single dose of ferric ions when compared with mice pretreated only with bLF, and the difference ranged from 25 to 35% on day-30.

Denisova et al. (1996) reported a synergistic effect of bLF, lactoperoxidase, and lactoglobulin obtained from cow milk, as well as hLF obtained from human milk, in protection against mice infected intranasally with *Shigella sonnei*. The combined administration of these preparations in doses, each one having no protective action, contributed to the elimination of the bacteria from the lungs and prevented the death of the animals. This phenomenon was found to be nonspecific and was suggested for use in the development of preparations for passive immunization against a wide spectrum of microorganisms.

Oral administration of bLF with milk in mice could suppress the proliferation of *Clostridium ramosum* C1 *in vivo* and decreased the numbers of *C. ramosum* and other bacteria in the feces (Teraguchi et al., 1995a). This bacteriostatic effect of bLF was depen-

dent on the concentration of bLF, the duration of feeding, and the administered dose of *C. ramosum* C1. Compared with bovine serum albumin, ovalbumin, bovine whey protein isolate, or bovine casein, only bLF showed this specific activity. A similar effect of bLF was observed after oral inoculation with *C. ramosum* JCM 1298, *C. paraputrificum* VPI 6372, or *C. perfringens* ATCC 13124. A hydrolysate prepared by digestion of bLF with porcine pepsin showed the same inhibitory effect on proliferation of *C. ramosum in vivo* as occurred with undigested bLF. These results indicate that ingested bLF could exert a bacteriostatic effect against clostridia in the gut even after enzymatic digestion in gut.

Furthermore, supplementation of the milk diet with bLF or a pepsin-generated hydrolysate of bLF also resulted in significant suppression of bacterial translocation from the intestines to the mesenteric lymph nodes, and the bacteria involved were mainly members of the family *Enterobacteriaceae* (Teraguchi et al., 1995b). This ability of LF to inhibit bacterial translocation may be due to its suppression of bacterial overgrowth in the guts of milk-fed mice.

Wada et al (1999) suggested that bLF exerts an *in vivo* inhibitory effect on colonizing *Helicobacter pylori* by detaching the bacterium from the gastric epithelium and by exerting a direct anti-bacterial effect. Germfree BALB/c mice were orally inoculated with *H. pylori* to induce infection. Three weeks after infection the mice were given bLF orally once daily for 2 or 4 weeks and were then killed to examine the bacterial number in the stomach and the serum antibody titer to *H. pylori*. To count the number of epithelium-bound *H. pylori*, the resected stomach was agitated in phosphate-buffered saline to remove non-bound *H. pylori* before bacterial enumeration. The administration of 1% bLF for 3 to 4 weeks decreased the number of *H. pylori* in the stomach to one-tenth and also exerted a significant inhibitory effect on the attachment of *H. pylori* to the stomach. Furthermore, the serum antibody titer to *H. pylori*, decreased to an undetectable level.

2. Viruses. Lu et al. (1987) examined the *in vivo* effects of holo-hLF on the survival rates of mice and titers of spleen focus-forming viruses (SFFV) in mice inoculated with the polycythemia-inducing strain of the Friend virus complex (FVC-P). LF prolonged the survival rates and decreased the titers of SFFV in mice infected with FVC-P. Titers of SFFV, assayed 14 days after administration of FVC-P, were measured by the spleen focus-forming unit assay in secondary mouse recipients. Decreases in titers of SFFV were apparent when LF was given *in vivo* as a single bolus dose of 200 μg within 2 h of the FVC administration, or as a total dosage of 200 μg given on days 1, 2, 4, 7, 9, and 11 after FVC-P, and to a lesser degree when LF was given as a total dosage of 200 μg on days 3, 4, 7, 9, and 11 after FVC-P. No decreases in titers of SFFV were detected when LF was given up to 3 days before or more than 3 days after FVC-P. LF did not appear to be directly inactivating the viruses as it did not inactivate the SFFV or the Friend murine leukemia helper virus *in vitro*. The results suggested that the protective effect of LF *in vivo* was probably due to an action on cells responding to the FVC or to an action on cells which influence the cells responding to the FVC or which influence the virus.

LF could inhibit bacterial growth in the murine conjunctival sac (Fujihara & Hayashi, 1995). The antiviral activity of LF against herpes simplex virus type-1 (HSV-1) *in vitro* (infectivity of Vero cell monolayers) and inhibit infection *in vivo* in the mouse cornea was also reported. LF effectively prevented HSV-1 plaque formation.

Administration of topical 1% LF prior to the virus inoculation suppressed
ocular tissue; however, it did not inhibit propagation of the virus.

The administration of bLF (1 mg/g body weight) before
cytomegalovirus (MCMV) infection completely protected the BALB/c mi
due to the infection (Shimizu et al., 1996). The LF-treated mice showed
increase in the NK cell activity but not of the cytolytic T lymphocytes tha
MCMV-derived peptide. Moreover, the elimination of the NK cell activity
with anti-asialo GM1 antibody abrogated such augmented resistance, thus
hypothesis that the LF-mediated antiviral effect *in vivo* is performed thr
mentation of NK cell activity. No such LF-mediated antiviral effect *in vivo* with the increased NK cell activity was found in athymic nude mice, whereas it was restored completely by the transfer of splenic T cells from LF-treated donors. These findings suggest that T lymphocytes induce both the augmentation of NK cell activity and the resultant antiviral effect in the LF-treated hosts.

Effects of oral administration of bLF on intractable stomatitis in feline immunodeficiency virus (FIV)-positive and FIV-negative cats, and phagocytosis of neutrophils in healthy and ill cats was tested (Sato et al., 1996). Bovine LF (40 mg/kg ot body weight) was applied topically to the oral mucosa of cats with intractable stomatitis daily for 14 days and improvement of clinical signs of disease (pain-related response, salivation, appetite, and oral inflammation) was evaluated. Assay of neutrophil phagocytosis was examined before and 2 weeks after starting LF treatment, using nonopsonized hydrophilic polymer particles. Bovine LF could improve intractable stomatitis and concurrently enhanced the host defense system. Topical application of bLF to oral mucus membrane was suggested useful as a treatment for intractable stomatitis also for FIV-positive cats.

3. Mycoplasma. The protective effect of LFcin against *Toxoplasma gondii* infection was examined in experimental murine toxoplasmosis (Isamida et al., 1998). All the mice that were dosed orally with LFcin (5.0 mg), and challenged with cysts of *T. gondii* at a dose of LD_{90} survived until the end of the experiment (35 days post challenge). Intraperitoneal administration of LFcin (0.1 mg) also prevented death in 100% of treated mice challenged with *T. gondii* cysts. In contrast, 80% of untreated mice died of acute toxoplasmosis within 14 days post challenge. In the mice treated per-orally with LFcin, the number of cysts in the brain was significantly lower than that in untreated mice. Levels of IFN-g in the serum of infected mice treated per-orally with LFcin showed a tendency to lower than those in the infected mice without treatment.

B. Immuno-modulatory functions

Natural killer (NK) and lymphokine-activated killer (LAK) cell cytotoxic functions are strongly augmented by LF (Shau et al., 1992). LF significantly enhances NK and LAK activities when added at the beginning of NK or LAK cytotoxicity assays. LF is effective in augmenting cytotoxic activities at concentrations as low as 0.75 μg/ml, and higher concentrations of LF induce greater augmentation of NK and LAK. Iron does not appear to be essential for LF to increase NK and LAK, as depleting iron from LF with the chelator deferoxamine does not affect the capacity of LF to increase cytotoxicity. LF is known to have RNase enzymatic activity and LF enhancement of NK and LAK can be blocked by RNA. However, LFs from two different sources with over 100-fold difference in RNase activity are equally effective in enhancing NK and LAK. Furthermore, purified

non-LF RNase does not modulate NK or LAK activity and DNA is as effective as RNA in blocking LF augmentation of NK or LAK cytotoxicity. Therefore, the RNase activity is unlikely to be responsible for LF enhancement of the cytotoxicity. Newborn infants are known to have low NK activity and NK and LAK cells have been implicated in host defense against microbial infections. Thus, maternal milk- derived LF may have a role in boosting antimicrobial immunity in the early stages of life. In adults, LF released from neutrophils may enhance NK and LAK functions in the inflammatory process induced by microbial infections.

Debbabi et al. (1998) investigated the nature of immune responses induced by repeated oral administration of bLF to mice (1998). Mice were fed daily for 4 weeks with two doses of protein antigen: a low (0.05 mg/g body weight per d) or high (1 mg/g body weight per d) dose of LF, or water as a control. A fourth group was immunized intramuscularly with 0.01 mg LF in complete Freund's adjuvant. Anti-LF IgA and IgG were detected in the intestinal fluid and serum of mice given LF. Total immunoglobulins were higher in the intestinal fluid in LF groups than in the control group. No difference could be detected in the serum. IgA and IgG secretion was enhanced in Peyer's patches and spleen from LF-fed mice, in comparison with controls. [^{3}H]thymidine uptake into Peyer's patch and spleen cells from both control and LF-fed mice was enhanced by 75 μg of LF/ml *in vitro*, but LF groups had a greater proliferation rate than the control group. These findings suggested that LF could act as an immunostimulating factor on the mucosal immune system and that activation of the mucosal immune system is dependent on the ability of LF to bind to the intestinal mucosa.

In vivo administration of bLF seems to inhibit delayed type hypersensitivity (DTH) reactions in mice (Zimecki & Machnicki, 1994). Administration of bLF at 48 or 24 h before eliciting the DTH reaction was not effective; however, bLF suppressed the reaction when given at the peak of the inflammatory process. The effects of bLF were strongest when the protein was injected intravenously. Intraperitoneal or intramuscular administrations of bLF were less inhibitory. In addition, bLF diminished, although to a much lesser degree, the inflammatory reactions induced by BCG antigen. The inhibitory action of bLF does not involve liver since treatment of mice with galactosamine does not reverse the inhibition. Studies on cytokine production revealed that peritoneal macrophages, derived from mice pretreated with LF, have an increased ability to produce IL-6 *in vitro* after induction with LPS. In addition, the inhibition of macrophage migration, mediated by migration inhibition factor, was abolished by bLF. However, the inhibitory effect of bLF could not be transferred with serum from donors treated with bLF.

Miyauchi et al. (1997) evaluated the immunomodulatory effects of bLF hydrolysate on the proliferation of murine splenocytes. The hydrolysate enhanced [^{3}H]thymidine uptake by splenocytes, but undigested bLF exerted an inhibitory effect. The hydrolysate had the ability to inhibit the blastogenesis induced by mitogens such as concanavalin A, phytohemagglutinin, and LPS; inhibition was similar to that with undigested bLF. These results suggested that the hydrolysate contained both immunostimulatory and immunoinhibitory peptides. The stimulatory effect of the hydrolysate in the absence of mitogens was then explored in more detail using nonadherent splenocytes. The proliferative response of splenocytes to the hydrolysate was much greater in the fraction that was enriched with B cells than in the fraction that was enriched with T cells. The hydrolysate did not affect thymocyte proliferation. These data indicated that the adherent cells resembling macrophages found among the splenocytes were not the target cells of

the hydrolysate. The stimulatory effect of the hydrolysate was due to the activation of B cells by the hydrolysate and enhanced immunoglobulin production by splenocytes. Because the hydrolysate also enhanced the proliferation and immunoglobulin A production of Peyer's Patch cells, the immunostimulatory effect of the hydrolysate *in vivo* was examined using mice that had been orally immunized with cholera toxin. The concentrations of immunoglobulin A conjugated against cholera toxin in bile and in the intestinal contents of mice fed liquid diets containing 1% (wt/vol) LF hydrolysate were greater than those of mice fed control diets. These results suggest a beneficial effect of LF on the mucosal immunity.

Zimecki et al. (1998a) also investigated the effect of bLF oral treatment on carrageenan-induced inflammation in rats. Animals were given 5 oral doses of bLF (10 mg each) on alternate days and 24 h after the last dose a carrageenan inflammation was induced in the hind foot. Control rats were given 0.9% saline or BSA. The magnitude of the reaction was measured after 2 h (optimal response) and expressed as an increase of the foot pad thickness in millimeters. The evaluation of bLF effects on carrageenan reaction was supplemented by determination of the ability of spleen cell cultures to produce IL-6 and TNF-α upon LPS induction. Bovine LF inhibited the carrageenan-induced inflammation in by 50 and 40% as compared to saline and BSA control groups, respectively. The inhibition was also associated with a substantial decrease in the ability of splenocytes to produce IL-6 in bLF-treated rats (94 and 83% as compared to saline- and BSA-treated groups). The LPS-induced TNF-α production was also decreased, although to a lesser degree (48 and 35%, respectively). The decreased ability of spleen cells to produce inflammatory cytokines in bLF-treated rats indicates that hypo-reactivity of the immune system cells may be the basis for the inhibition of carrageenan-induced inflammation.

Intravenous (i.v.) administration of bLF (10 mg/mouse), 24 h before thymectomy, reduced, on average, the level of serum IL-6 by 70% as measured 4 h after surgical procedure (Zimecki et al., 1998b). The inhibiting effect of bLF on TNF-α production was low with a mean 30% reduction. The effects of bLF (i.v.) administration on the cytokine levels following splenectomy were less inhibitory. Bovine LF caused an approximate 35% fall in IL-6 levels and even weaker effects (20% inhibition) on TNF-α release. Application of much lower (1-0.2 mg) doses of bLF was even more effective in lowering IL-6 levels after thymectomy (up to 90%) after 5 bLF doses, and by 55% of TNF-α. The authors suggest that LF may find therapeutic application for diminishing manifestations of shock caused by clinical insults.

Zimecki et al. (1998c) conducted a clinical trial in Poland to monitor several immune parameters in 17 healthy volunteers taking commercially available capsules containing bLF (40 mg/ daily) orally for 10 days. Leukocyte number and content of main blood cell types, spontaneous and phytohemagglutinin A (PHA)-induced proliferation of lymphocytes, plasma levels of IL-6 and TNF-α, as well as spontaneous and LPS-induced production of these cytokines in peripheral blood cell cultures, were evaluated. All measurements were performed before, one day and 14 days following cessation of bLF treatment. A transient drop was observed in the percentage of neutrophils accompanied by an opposite phenomenon with regard to lymphocyte levels. More profound changes were registered in the percentage of other cell types, such as a 100% increase in the level of immature cell forms. At the same time the percentages of eosinophils and monocytes markedly declined. All these changes were, however, more individual and regulatory, the direction of these changes depended on initial profile of blood cells. Although the prolif-

erative response of lymphocytes showed, on average, a transient decrease, differentiated effects of bLF treatment were observed depending on initial ability of lymphocytes to proliferate. TNF-α serum levels showed a tendency to decrease during the monitoring time, the changes of IL-6 levels were, however, not significant. As in the case of the proliferative response, the treatment with bLF was regulatory with respect to serum TNF-α levels. The influence of other ingredients such as selenium or vitamins, contained in the capsules, cannot be excluded, although the data indicated that orally taken bLF alone can induce identical changes as the capsules containing bLF. The study suggests that oral administration of bLF-containing capsules may regulate certain immune responses in healthy individuals.

In a clinical trial conducted in Japan, Yamauchi et al. (1998) examined the effects of orally administered bLF on the immune system of healthy volunteers. Ten healthy male volunteers (age range: 31-55) were given bLF (2 g/body/day) for 4 weeks. Blood samples were drawn before, during and after LF administration. Phagocytic activity and superoxide production activity of PMN were evaluated from the number of polymer particles phagocytosed by PMN and by the dichlorofluorescein (DCFH) oxidation assay, respectively. The expression levels of CD11b, CD16 and CD56 molecules on leukocytes were quantified using flow cytometry. The phagocytic activity of PMN increased during the period of LF administration in 3 of the 10 volunteers. In 2 of the 3 volunteers in which the phagocytic activity increased, PMN expressed CD16 at higher levels corresponding to the increase in 3 of the 10 volunteers, whereas the CD11b+ lymphocytes and CD56+ lymphocytes increased in 4 volunteers including the same 3 volunteers who showed an increase in CD16+. These results suggest that the proportion of NK cells among the lymphocytes might have increased in these subjects. It was demonstrated that the phagocytic activity or superoxide production activity of PMN or the proportions of CD11b+, CD16+ and CD56+ in lymphocytes was influenced by LF administration in 7 of the 10 volunteers, while the effects of LF on the immune system differed in individual cases. These results suggest that LF administration may influence primary activation of the host defense system.

C. Nutritional functions

LF has been suggested to play an important role in the intestinal absorption of iron (de Vet & van Vugt, 1971; de Vet & van Gool, 1974). During early life, infants usually consume a diet that is heavily dominated by milk. The bio-availability of the trace elements iron, zinc, copper and manganese from human milk is high compared to cow's milk and infant formulas. This high bio-availability may be explained by the presence of LF in human milk, which may facilitate iron and manganese uptake via an intestinal receptor for this protein (Lönnerdal, 1989).

1. Specific LF-binding receptors on intestinal brush border. Several lines of evidence have been suggested the occurrence of a specific LF receptor in the small intestinal brush-border membrane in several animal species, which is thought to be involved in LF-mediated intestinal iron absorption. Hu et al. (1990) isolated and partially characterized an LF receptor from mouse intestinal brush border. The receptor has been purified to homogeneity by affinity chromatography on an immobilized hLF column. The purified receptor was found to be active in that it binds apo- and holo-LF with a K_a of 0.1 μM. Anti-receptor antibodies were prepared, and the receptor was further isolated by

immunoaffinity chromatography in higher yield but in a denatured form. The purified receptor has a molecular mass of about 130-kDa, and consisted of a single polypeptide chain with an isoelectric point at 5.8. The receptor was also shown to bear concanavalin A and phytohemagglutinin binding glycans.

Kawakami and Lönnerdal (1991) have isolated a hLF receptor from solubilized human fetal intestinal brush-border membranes. The molecular weight of the receptor was 110-kDa by SDS-PAGE under non-reducing conditions and 37-kDa under reducing conditions. Competitive binding studies demonstrated specific binding of hLF. The binding was pH dependent, with an optimum between pH 6.5 and 7.5. Scatchard plot analysis indicated 4.3 x 10^{14} binding sites/mg membrane protein with an affinity constant of 0.3 μM for hLF. Both half-LF and deglycosylated LF bound to the receptor with an affinity similar to intact LF. In contrast, little binding of bLF or human TF to human brush border membrane vesicles occurred.

Rosa and Trugo (1994) investigated iron uptake from hLF by brush-border membrane vesicles (BBMV) obtained from the small intestine of human neonates. Uptake experiments were performed by incubation of ^{55}Fe-citrate or ^{55}Fe-LF with BBMV, followed by rapid filtration through microporous filters. ^{55}Fe uptake from LF by BBMV was dependent on pH, with a maximum at 7.5, and increased with incubation time, reaching a maximum at 1 min. When ^{55}Fe was bound to citrate, uptake was slower (maximum at 5 min) and not dependent on pH. In both experiments, the maximum uptake of iron bound to LF was about twice that of iron bound to citrate (230 pmol and 105 pmol/mg microvillus protein, respectively). Partial degradation of LF in two fragments resulted in the loss of its capacity to increase iron uptake by BBMV. The study suggested that LF could increase iron absorption during the neonatal period, contributing to the high bio-availability of this mineral in human milk.

Differentiated Caco-2 colon carcinoma cell monolayers grown in bicameral chambers have been used as an *in vitro* model to study the effect of different carrier molecules on mucosal iron transport (Sanchez et al., 1996). Transfer of iron across the monolayers in the apical-to-basolateral direction was greater from holo-LF than from iron citrate, while very little transport occurred from holo-TF. However, a greater proportion of iron was retained by the cells when Fe-citrate was the donor. Caco-2 cells expressed TF receptors (1.3 x 10^5/cell; K_a:20 nM), but binding of LF, though substantial in quantity, had an affinity too low to measure. When monolayers were incubated with ^{125}I-labelled LF or TF some ^{125}I-activity was transported, but almost all was TCA-soluble, suggesting that degradation products rather than intact protein were being transported. Addition of 10 μM S-nitroso-N-acetyl-D,L-penicillamine, which produces nitric oxide (NO) in solution, caused a significant increase in iron transport from ferric citrate, but not from Fe-LF or Fe-TF. It was concluded that in this *in vitro* system LF but not TF enhances mucosal iron transport, and that NO may play a regulatory role in iron absorption.

2. LF-mediated intestinal iron uptake. Various clinical trials and *in vivo* animal experiments on the role of LF in intestinal iron absorption are inconclusive. The efficacy of supplementing iron bound to LF to iron-deficient and iron-sufficient young mice was evaluated in comparison with supplementation of iron as iron chloride (Fransson et al., 1983). Mice fed a non-supplemented milk diet (approximately 1 mg Fe/L) for 4 weeks had a microcytic, hypochromic anemia and low tissue iron concentrations. Iron supplementation of the diet with LF-iron, or iron chloride at a level of 5 mg Fe/L prevented the ane-

mia and resulted in tissue iron levels similar to levels found for mice fed a stock commercial diet. There was no significant difference in any of the parameters analyzed between the groups of mice receiving the two iron supplements following a diet deficient in iron. Apo-LF when supplemented to the diet had no negative effect on the iron status of the mice. These results suggested that LF could be a useful vehicle for supplementation of iron.

Davidson et al. (1990) examined the iron retention from human milk, milk-based infant formula (IF) with and without supplemental ferrous sulfate, and IF supplemented with either hLF or bLF in infant rhesus monkeys. There was no significant difference in iron retention from the experimental diets: human milk, IF, IF+Fe, IF+hLF, or IF+bLF. The authors concluded that the infant monkeys absorb and retain iron similarly from human milk and infant formula. Supplementation of infant formula with hLF or bLF resulted in similar iron retention to that of ferrous sulfate-supplemented infant formula.

Iron balance studies were performed in 16 term infants from week-3 through week-17 of life (Schulz-Lell et al., 1991). The balance studies were performed at home and comprised five periods with an interval of 3 to 4 weeks, each consisting of three 24-h collections of milk and stool samples. Seven infants were fed an adapted infant formula supplemented with bLF (1 mg/ml) and nine received the same formula without bLF. The LF supplemented group received 169 μg iron/kg b.w/day and retained 63 μg/kg b.w./day. The mean iron intake of infants fed with the adapted formula without supplementation of bLF was 118 μg/kg b.w./day. The retention of iron was 43 μg/kg b.w./day. Mean percentage retention of iron in the supplemented group was 36%, in the non-supplemented group 28%.

In a study by Chierici et al. (1992) infant formulae supplemented with various amounts of bLF were given to two groups of infants. These infants were compared with infants receiving unsupplemented formula and breast-fed infants. The effects of these diets on levels of hemoglobin, hematocrit, serum iron, ferritin and zinc were examined for a study period of 150 days. At birth, concentrations of iron, hemoglobin, hematocrit and zinc were comparable in all four feeding groups. The serum zinc level was not altered by LF supplementation. Ferritin levels of breast-fed infants were significantly higher than in non-supplemented formula-fed infants at day 30 and day 90. This difference was seen only at day 30, when comparing breast-fed infants to LF-supplemented formula-fed infants. Comparing the infants receiving formulae, the formula supplemented with the higher amount of bLF induced significantly higher serum ferritin levels compared to the unsupplemented formula at day 90 and day 150. These observations favor the hypothesis that LF is possibly involved in iron absorption.

The following studies, however, do not support the efficacy of LF in facilitating iron uptake in the gastrointestinal tract. McMillan and co-workers (1977) compared iron availability from human milk with that from other formulas supplemented with LF. The study concluded that manipulation of the protein, fat, lactose, calcium, phosphorus, or LF content of proprietary milk did not reproduce the iron absorption demonstrated with human milk. Studies by Fairweather-Tait et al. (1987) with ^{59}Fe-labels suggested that bLF has no effect on iron absorption in rats. The results were compared with those obtained from a group of infants fed a similar level of iron as ferric chloride, labeled with ^{59}Fe, together with 30 mg ascorbic acid. There was a very wide variation in percent iron retention amongst the infants but no overall difference between the LF and ferric chloride groups.

The potential effect of LF on iron absorption was investigated by measuring iron absorption in infants fed breast milk (with its native content of LF) and the same milk from which LF had been removed (> 97%) by treatment with heparin-Sepharose. Eight breast-fed infants (2-10 mo; mean age 5 mo) were fed 700 to 1000 g of each milk in a randomized, cross-over design with each child acting as his/her own control. The milk was labeled with 8.6 μM (0.5 mg) of ^{58}Fe and iron absorption was measured by quantifying the incorporation of the isotope into red blood cells 14 d after intake using thermal ionization mass spectrometry. Fractional iron absorption was significantly lower from breast milk than from LF-free breast milk. The geometric mean (range) was 11.8% (3.4-37.4%) for breast milk and 19.8% (8.4-72.8%) for LF-free breast milk. These results do not support a direct role for LF in the enhancement of iron absorption from human milk at this age. In addition, iron absorption (11.8%) from human milk fed over several feeds was lower than that previously reported for single feed studies.

D. Physiological functions

The germfree, colostrum-deprived, immunologically 'virgin' piglet model was used to evaluate the ability of bLF to protect against lethal shock induced by intravenously administered endotoxin (Lee et al., 1998). Piglets were fed bLF or BSA prior to challenge with intravenous *E. coli* LPS, and temperature, clinical symptoms, and mortality were observed for 48 h following LPS administration. Prefeeding with LF resulted in a significant decrease in piglet mortality compared to feeding with BSA (16.7 versus 73.7% mortality). Protection against the LPS challenge by LF was also correlated with both resistance to induction of hypothermia by endotoxin and an overall increase in wellness, as quantified by a toxicity score developed for these studies. *In vitro* studies using a flow cytometric assay system demonstrated that LPS binding to porcine monocytes was inhibited by LF in a dose-dependent fashion, suggesting that the mechanism of LF action *in vivo* may be inhibition of LPS binding to monocytes/ macrophages and, in turn, prevention of induction of monocyte/macrophage-derived inflammatory-toxic cytokines.

Antibody to an estrogen inducible mouse uterine protein (Teng et a., 1986) has been used to isolate cDNA to the messenger RNA. Analysis of the deduced primary structure and additional biochemical characterization indicated that the protein is LF. An increase in the level of LF mRNA of at least 300-fold can be induced in the mouse uterus by estrogen (Pentecost & Teng, 1987).

Neutrophils can inactivate LPS and block its ability to prime fresh neutrophils for enhanced fMLP-triggered release of superoxide. Wang et al. (1995) showed that the inactivation of LPS by neutrophils was primarily due to LF. A time course for inactivating LPS showed that neutrophils (5 million/ml) took 30 min to inactivate 10 ng/ml LPS. Mononuclear cells could not inactivate LPS under the same conditions. Experiments with radioactive LPS showed that inactivated LPS remained in the medium and was not taken up or destroyed by the neutrophils during inactivation. Inactivated LPS still gelled Limulus lysate and primed monocytes. Cell-free medium from neutrophil suspensions also inactivated LPS. A single LPS-inactivating factor was purified from medium by heparin-agarose chromatography. SDS-PAGE showed a single band at 80-kDa, which was identified as LF by immunoblotting. Anti-LF immunoglobulin G removed the LPS-inactivating activity from purified LF and cell-free medium. Purified neutrophil LF required 30 min to inactivate LPS, indicating inherently slow binding of LF to LPS.

Burrin et al. (1996) examined the anabolic effect of orally administered bLF on visceral organ growth and protein synthesis in newborn pigs. We studied a total of 18 unsuckled newborn pigs. Pigs were randomly assigned to one of three dietary treatment groups: bottle-fed (10 ml/h) formula, formula containing physiologic levels (1 mg/ml) of added bLF, or colostrum. After 24 h of feeding, the visceral organ protein synthesis *in vivo* was measured using a flooding dose of [^{3}H]phenylalanine. The visceral organ protein and DNA mass, as well as intestinal hydrolase activities and villus morphology was also measured. Hepatic protein synthesis in pigs fed either formula containing bLF or colostrum was similar and in both groups was significantly higher than in pigs fed formula. Splenic protein synthesis was not significantly different in pigs fed either formula or formula containing bLF, but was significantly higher in colostrum-fed animals. There were no significant differences in small intestinal growth, protein synthesis, or hydrolase activities between newborn pigs fed formula, formula containing bLF, or colostrum. These results indicate that feeding formula containing physiologic concentrations of added bLF could increase hepatic protein synthesis in newborn pigs, suggesting that colostrum-borne LF serves an anabolic function in neonates.

Antitumor activity. Yoo et al. (1997) studied the effect of bLF and bLFcin on inhibition of metastasis in murine tumor cells, B16-BL6 melanoma and L5178Y-ML25 lymphoma cells, using experimental and spontaneous metastasis models in syngeneic mice. Subcutaneous (s.c.) administration of apo-bLF (1 mg/mouse) and bLFcin (0.5 mg/mouse) 1 day after tumor inoculation significantly inhibited liver and lung metastasis of L5178Y-ML25 cells. However, apo-hLF and holo-bLF at the dose of 1 mg/mouse failed to inhibit tumor metastasis of L5178Y-ML25 cells. Similarly, the s.c. administration of apo-bLF as well as bLFcin, but not apo-hLF and holo-bLF, 1 day after tumor inoculation resulted in significant inhibition of lung metastasis of B16-BL6 cells in an experimental metastasis model. Furthermore, *in vivo* analysis for tumor-induced angiogenesis, both apo-bLF and bLFcin inhibited the number of tumor-induced blood vessels and suppressed tumor growth on day 8 after tumor inoculation. However, in a long-term analysis of tumor growth for up to 21 days after tumor inoculation, single administration of apo-bLF significantly suppressed the growth of B16-BL6 cells throughout the examination period, whereas bLFcin showed inhibitory activity only during the early period (8 days). In spontaneous metastasis of B16-BL6 melanoma cells, multiple administration of both apo-bLF and bLFcin into tumor-bearing mice significantly inhibited lung metastasis produced by B16-BL6 cells, though only apo-bLF exhibited an inhibitory effect on tumor growth at the time of primary tumor amputation (day-21) after tumor inoculation. These data suggest that apo-bLF and bLFcin inhibit tumor metastasis through different mechanisms, and that the inhibitory activity of bLF on tumor metastasis may be related to iron-saturation.

The influence of bLF on colon carcinogenesis was investigated in male F344 rats treated with azoxymethane (Sekine et al., 1997a). After three weekly injections of azoxymethane, the animals received 2 or 0.2% bLF for 36 weeks. No effects indicative of toxicity were noted, but significant reduction in both the incidence and number of adenocarcinoma of the large intestine was observed with both doses. Thus, the incidences of adenocarcinoma in the groups receiving 2% and 0.2% bLF were 15% and 25%, respectively, in contrast to the 57.5% control value. These results suggest a possible application for bLF in the chemoprevention of colon cancer. Sekine et al. (1997b) also reported that

the administration of bLF (2%) and *Bifidobacterium longum* (3%) significantly decreased the numbers of aberrant crypt foci. Most importantly large size foci composed of four or more crypts were always significantly decreased by 2% bLF. Studies on the natural killer activity of spleen cells demonstrated enhancement by bLF and *B. longum* in line with the levels of influence on foci induction, indicating a possible role for elevated immune cytotoxicity in the observed inhibition.

Iigo et al. (1999) reported that oral administration of bLF and the bLF hydrolysate also demonstrated significant inhibition of lung metastatic colony formation from s.c. implanted tumors without appreciable effects on tumor growth. bLFcin displayed a tendency for inhibition of lung metastasis. On the other hand, bLF did not exert marked anti-metastatic activity in athymic nude mice bearing Co 26Lu, though bLF had a tendency to inhibit the lung metastatic colony formation associated with anti-asialoGM1 antibody treatment. AsialoGM1+ and CD8+ cells in white blood cells were increased after treatment with bLF. *In vitro*, the viability of Co26Lu-F55 cells was markedly decreased when co-cultured with white blood cells from mice administrated bLF, but recovered on treatment with anti-asialoGM1 antibody or anti-CD8 mAb and complement. The results suggest bLF and related compounds might find application as tools in the control of metastasis and that asialoGM1+ and CD8+ cells in the blood are important for their inhibitory effects.

X. APPLICATIONS

The emerging knowledge about diseases and the role of natural foods in lowering the risk of such disease processes, and research efforts to identify and develop nutritionally important supplemental foods have been steadily increasing. Milk contains the widest range of biologically active ingredients. LF is one such bioactive protein present in milk (milk-derived whey or derivatives). It is a complex molecule with a number of potential functional properties (Naidu & Bidlack, 1998).

A. Potential additive for food safety

Consumption of cow's milk has been an integral part of human civilization since antiquity. LF is now recognized as a significant constituent of milk, responsible for nutraceutical benefits and innate protection against intestinal illnesses. A vast amount of literature has been published on the prophylactic, therapeutic, and regulatory role of LF in various physiological functions during the past four decades. The following points validates milk LF as a potent natural antimicrobial for food safety.

Ultimately, an antimicrobial must be non-toxic to test animals and humans. It is also important that the antimicrobial be metabolized and excreted by the body. *In vivo* studies have demonstrated that the liver plays a central role in the elimination of LF from plasma. When injected by intravenous route, LF is taken up rapidly by the receptors in the liver reticuloendothelial system. The integrity of the protein moiety of LF is required for its effective uptake by the liver. Histological and cytochemical investigations indicate that sinusoidal and Kupfer cells are responsible for hepatic uptake of LF (Courtoy et al., 1984). Studies with iodinated LF identified parenchymal cells as the prime source of LF catabolizing activity (Ziere et al., 1992). The liver degrades LF with a half-life of 2.7 h (Regoeczi et al., 1985). LF seems to remain associated with the plasmalemma for prolonged periods of time (Ziere et al., 1992) which could be interpreted as a sign of slow

internalization. However, LF is readily detected in both endosomes and lysosomes as early as 15 min after administration (Regoeczi et al., 1985). LF appears in bile 35 min after intravenous injection (Regoeczi et al., 1994).

Water solubility of an antimicrobial agent is probably the most important physical property. Hydrophilic properties are necessary to assure that the antimicrobial agent is soluble in aqueous phase where microbial growth occurs. LF is water-soluble, yet dissolves in a wide range of inorganic as well as organic solvents.

Thermal behavior can also directly influence the activity of an antimicrobial agent, especially with its carry-through properties. If the food is heated during processing, a highly volatile agent such as a phenolic compound can be vaporized and lost. Paulsson et al. (1993) examined the heat-induced enthalpy changes in apo- and iron-saturated bLFs by differential scanning calorimetry. Two thermal transitions with varying enthalpies were observed, depending on the iron-binding status of the protein. Iron-saturated LF was more resistant to heat-induced changes than was the apo-LF. Native LF had two transitional peaks and pasteurization affected only the low temperature transition. Iron-saturated LF revealed a single transitional peak that was resistant to denaturation by pasteurization temperatures. Native LFs, both unheated and pasteurized, showed similar antibacterial properties suggesting temperature-resistance of the molecule.

The ability of an antimicrobial compound to ionize can alter its biological activity depending on the pH of the food system in which it is used. LF exerts antimicrobial activity at a wide pH range in different ionic milieus. Natural compounds, occurring as the major food components, can also influence activity of an antimicrobial depending on the properties of the antimicrobial and those of the naturally occurring compound. The cationic nature enables LF to interact and complex with other proteins such as casein, α-lactalbumin (Hekman, 1971; Lampreave et al., 1990), lysozyme (Perraudin et al., 1974), and immunoglobulin-A (Watanabe et al., 1984). Various regulatory functions were attributed to such LF complexes. Interactions with acidic polysaccharides (Mann et al., 1994) which are ubiquitous components of cell membranes. The arginine-rich N-terminus of the LF molecule (Mann et al., 1994) allows LF to bind with varying degrees of affinity to almost any cell, as well as to low-density lipoproteins.

Knowledge of the mode of action of an antimicrobial and of the ability of the organisms to overcome this mode of action can be helpful in determining the efficiency and usefulness of an antimicrobial. The protective role of LF in host defense against various microbial illnesses has been established (Naidu & Arnold, 1997). This chapter has identified several of the molecular mechanisms related to LF antimicrobial activity.

Valid assay methods for antimicrobial agents are essential so that the levels can be easily determined. Various laboratories have developed rapid, sensitive, enzyme-linked immunosorbant assays (ELISA) for quantitation of LF in different biological systems. Shinmoto et al. (1997) reported a highly sensitive competitive ELISA for bLF. Culliere et al. (1997) recently developed a microparticle-enhanced nephlometric immunoassay for LF quantitation. This assay is sensitive (detection limit in reaction mixture, 0.2 mg/L) and can be performed in diluted milk (1/300 in reaction mixtures), excluding any interference or sample pretreatment. It allows the quantification of LF on a large range of concentrations (0.675-21.6 g/L) with accuracy (linear-recovery in dilution-overloading assay) and precision (3-6% variation). A rapid immunoluminometric assay and a non-competitive avidin-biotin immunoassay were also developed for LF quantitation. The LactoCard, a new commercial assay, allows rapid determination of LF concentrations in 10-15 min.

TABLE 7. In vivo effects of LF - human clinical trials and consumption studies in different experimental animal models.

Test parameter	Model	Test design (Treated/Total)	LF	Dosage	Carrier	Route	Reference
Effect on intestinal iron absorption	Mouse	Suckling mice	bLF	4 g/L or 50 µM for 4 wks	Milk	Oral	Fransson et al., 1983
Effect on tissue mineral content	Mouse	Suckling mice	bLF	4 g/L or 50 µM for 4 wks	Milk	Oral	Keen et al., 1984
Prophylaxis during neutropenia	Human	Cancer patients (5/14)	hLF	800 mg/day for 10 d	Capsule	Oral	Trumpler et al., 1989
Effect on fecal microflora	Human	Newborn term infants (38/69)	bLF	2.8 g/L for 14 d	Formula	Oral	Balmer et al., 1989
Iron absorption in the newborn	Monkey	Newborn monkeys (6/6)	bLF	1 g/L one time per diet	Formula	Oral	Davidson et al., 1990
Effect on muscle growth	Rats	Adult animals (48/80)	bLF	0.25, 1 or 4 mg/d for 14-d	Saline	S.C.	Byatt et al., 1990
		Neonatal animals (40/60)	bLF	0.01-0.1 mg/kg b.wt.	Saline	S.C.	Byatt et al., 1990
Effect on iron balance	Human	3-week old infants (7/16)	bLF	1 g/L for 3.5 mo.	Formula	Oral	Schulz-Lell et al., 1991
Effect on fecal microflora	Human	New born term infants (29/55)	bLF	0.1 or 1.0 g/L for 3 mo.	Formula	Oral	Roberts et al., 1992
Serum Zn, Fe & ferritin levels	Human	New born term infants (28/51)	bLF	0.1 or 1.0 g/L for 5 mo.	Formula	Oral	Chierici et al., 1992
Control of *M. paratuberculosis*	Mouse	Gnotobiotic mice (8/32)	bLF	2 mg/L for 10 mo.	Water	Oral	Hamilton & Czuprynski, 1992
Effects on fecal flora	Mouse	Gnotobiotic mice (5/75)	bLF	2 g/L for 14 days	Formula	Oral	Hentges et al., 1992
Effects on induced pancreatitis	Human	Adult subjects (6/35)	bLF	100 mg/kg b. wt./ dose	Saline	I.P	Koike & Makino, 1993
Effect on iron balance	Bovine	New born calves (7/53)	bLF	11.4 g /day for 5 d	Colostrum	Oral	Kume et al., 1994
Effect on iron balance	Bovine	1-week old calves (12/36)	bLF	5 g/day for 5-10 d	Formula	Oral	Kume et al., 1995
Antitumor activity	Human	Adult cancer patients (7/7)	bLF	0.7% per day for 6 mo.	Formula	Oral	Kennedy et al., 1995
Bacteriostasis of *Clostridium spp.*	Mouse	(9/9); (15/25) & (30/40)	bLF	0.5 – 2.0% for 7-14 d	Milk	Oral	Teraguchi et al., 1995a
Effect on bacterial translocation	Mouse	5-week old mice (20/30)	bLF	0.5 – 2.0% for 7 d	Milk	Oral	Teraguchi et al., 1995b
Antiviral activity against stomatitis	Cats	Cats 5-10 yr. (1/7) (5/12)	bLF	40 mg/kg b.wt for 14 d	None	Oral	Sato et al., 1996
Protein synthesis *de novo*	Porcine	New born pigs (6/18)	bLF	1 g/L for 24 h	Formula	Oral	Burrin et al., 1996
Inhibition of induced colon tumors	Rats	6-wk rats (70/115)	bLF	0.2 to 2.0% for 30 wks	Diet	Oral	Sekine et al., 1997a
Inhibition of induced colon tumors	Rats	6-wk rats (16/30) (8/15)	bLF	0.2 to 2.0% for 4-13 wks	Diet	Oral	Sekine et al., 1997b
Control of *Toxoplasma gondii*	Mouse	8-wk mice (25/30)	LFcin	0.2 g/kg b.wt for 8-d	Saline	I.P	Isamida et al., 1998

XI. SAFETY AND TOLERANCE

LF is already used in a wide range of products including infant formulas, sport and functional foods, personal care products, as well as veterinary and feed specialties. In United States, LF is yet to be considered for a GRAS status. On the other hand, LF products have found a niche in various parts of Europe and southeast Asia controlled by the following regulations.

A. Legal status

In Europe, bLF is manufactured according to the Dairy Hygiene Directive 92/46 of the EC and meets the corresponding requirements. As per the EC regulations, bLF can be added to food for nutritional reasons. However, for allowed use, the relevant food standards should be considered.

In Japan, LF concentrates are allowed under number 438 as 'a substance composed mainly of LF obtained from mammal's milk' in food according to the Ministry of Health and Welfare Announcement No.160 of August 10, 1995.

According to the Public Code of Food Additives in South Korea, LF concentrates are allowed in food products. No special provisions are laid down.

In Taiwan, LF may be used in special nutritional food stuffs under the condition 'only for supplementing foods with an insufficient nutritional content and may be used in appropriate amounts according to actual requirements'.

B. *In vivo* metabolism and turnover

Bennett and Kokocinski (1979) measured the turnover of ^{125}I-LF in ten adult human subjects by simultaneous organ radioactivity counting procedure. Ferrokinetic studies were performed in three adults after the intravenous injection of [^{59}Fe]LF. 3. LF was rapidly eliminated from the plasma with a mean fractional catabolic rate of 5.7/day. Apo-LF (one subject) was eliminated at a slower rate (fractional catabolic rate 1.22/day). Of the administered ^{125}I label 99% was recovered in the urine, as free iodine, within the first 24 h. In the ^{59}Fe studies no appreciable activity was found in the urine. Organ radioactivity counting showed that LF was rapidly taken up by the liver and spleen. In the ^{125}I studies the rapid excretion of free ^{125}I suggested catabolism at these sites. In the ^{59}Fe studies, the radioactivity persisted in the liver and spleen for several weeks and was slowly transferred to the bone marrow before appearing in circulating erythrocytes. From the values of fractional catabolic rate, plasma LF, neutrophil LF and plasma volume, a 'derived neutrophil turnover' was calculated for each subject. The mean value was 8×10^8 neutrophils/day. This is about 1% the value obtained from the actual measurement of labeled cells. It is postulated that this 'derived value' represents only that portion of neutrophil turnover accounted for by intravascular senescence.

Ziere et al. (1992) characterized the hepatic recognition of LF. Intravenously injected ^{125}I-LF was cleared rapidly from the circulation by the liver (93% of the dose at 5 min after injection). Parenchymal cells contained 97% of the hepatic radioactivity. Internalization, monitored by measuring the release of liver-associated radioactivity by the polysaccharide fucoidin, occurred slowly. Only about 40% of the liver-associated LF was internalized at 10 min after injection, and it took 180 min to internalize 90%. Subcellular fractionation indicated that internalized LF is transported to the lysosomes.

Binding of LF to isolated parenchymal liver cells was saturable with a dissociation constant of 10 μM (20 x 10^6 binding sites/cell).

Regoeczi et al. (1994) studied the ability of liver to transfer LF from the plasma to the bile by injecting a dose (10 to 20 μg/100 gm) of labeled bLF intravenously and following its appearance in bile over 3 h. Both diferric- and apo-LF peaked in the bile 35 min after administration (i.e., the same time as bovine lactoperoxidase and diferric rat TF). However, only a small portion of the LF dose (approximately 1%) was recovered with the bile in 3 h. On the basis of autoradiographic evidence, the excreted LF appeared intact. The biliary excretion profile of albumin, a protein thought to reach the canaliculus by para-cellular diffusion, was notably devoid of a peak. This, together with competition observed between LF and lactoperoxidase on one hand and diferrin-TF and LF on the other for transfer to bile, suggests that LF is routed through the hepatocyte in vesicles. The process is initiated by binding to a plasma membrane component to which lactoperoxidase and diferric-TF can also bind. Most ^{59}Fe bound to LF accompanied the protein carrier to the bile. The study concluded that under normal circumstances (i.e., when concentration of LF in the plasma is very low), LF transferred from plasma by the liver is probably not the major source of this protein in bile.

C. Consumption studies and safety data

The *in vivo* efficacy as well as safety of LF has been documented in human clinical trials and in several consumption studies in experimental animal models (*Table 7*). Infant food formulae have been supplemented with bLF in four different clinical trials at levels of 0.1 to 1.0 g/L for three to five months and at 2.8 g/L for fourteen days to term human infants with no adverse effects. Also, bLF has been safety administered orally to cancer patients at 1.6 g/day for six months.

Much higher levels of bLF have been administered orally to mice and rats, as high as 20 g/L of milk for fourteen days and 20 g/kg diet for thirty weeks, respectively, with no known side effects. Subcutaneous or intraperitoneal administration of bLF also support the safety of bLF when it is given by more sensitive routes of delivery. Bovine LF has been given as a single intraperitoneal dose to rats at 100 mg/kg body weight with no known adverse effect. Other animal species, including cats, pigs, calves, and monkeys have been given bLF orally with no detrimental effects.

Bovine LF demonstrates structural homology and functional similarity to hLF. However, cow milk contains only about one-tenth the amount of LF compared to its occurrence in human milk. The levels of hLF in colostrum and mature milk is about 30 g/L and 2 g/L. Thus, a safe upper level of LF consumption by children, adolescents, and adults would be expected at these levels.

The literature describes no reports on toxicity of LF. No standard toxicology tests (acute, subchronic, chronic, carcinogenicity, reproductive, developmental, etc.) of LF were reported, Furthermore, LF has not been assigned a Chemical Abstract Services (CAS) registry number.

XII. SUMMARY

LF is an antimicrobial glycoprotein present in milk and various exocrine secretions that bathe the mucosal surfaces. This metal-binding protein is a multifunctional bioactive molecule with a critical role in many important physiological pathways. LF could

elicit a variety of inhibitory effects against microorganisms comprising stasis, cidal, adhesion-blockade, cationic, synergistic and opsonic mechanisms. Broad-spectrum activity against different bacteria, viruses, fungi and parasites, in combination with anti-inflammatory and immunomodulatory properties makes LF a potent innate host defense mechanism. The current global production of bovine milk LF is approximately 100 metric tons and this figure is continuously increasing. This protein is finding it applications as an active ingredient in infant formulae, and health foods in South-East Asian countries, in particular. LF is also in use as a therapeutic and prophylactic agent to control intestinal illnesses and mucosal infections. Certain oral hygiene products, skin care products, and animal feed supplements contain LF. Recent advances in LF research to elucidate the structural function relationships, antimicrobial mechanisms, cost-effective technologies for large-scale protein isolation and biotechnology is opening unlimited opportunities for this natural antimicrobial in the development of new products and formulation. A number of efficacy studies and clinical trials are ongoing in various laboratories with over 100 patents filed on this molecule in the past 10 years. Undoubtedly, LF is emerging as one of the leading natural microbial blocking agent in food safety and preservation.

XIII. REFERENCES

1. Abramson, J.S., Parce, J.W., Lewis, J.C., Lyles, D.S., Mills, E.L., Nelson, R.D., and Bass, D.A. 1984. Characterization of the effect of influenza virus on polymorphonuclear leukocyte membrane responses. *Blood* 64:131-138.
2. Aguas, A., Esaguy, N., Sunkel, C.E., and Silva, M.T. 1990. Cross reactivity and sequence homology between the 65-kilodalton mycobacterial heat shock protein and human lactoferrin, transferrin, and DR beta subsets of major histocompatibility complex class II molecules. *Infect. Immun.* 58:1461-1470.
3. Ahmadzadeh, N., Shingu, M., and Nobunaga, M. 1989. Iron-binding proteins and free iron in synovial fluids of rheumatoid arthritis patients. *Clin. Immunol.* 8:345-351.
4. Ainscough, E.W., Brodie, A.M., Plowman, J.E., Bloor, S.J., Loehr, J.S., and Loehr, T.M. 1980. Studies on human lactoferrin by electron paramagnetic resonance, fluorescence, and resonance Raman spectroscopy. *Biochemistry* 19:4072-4079.
5. Aisen, P., and Leibman, A. 1972. Lactoferrin and transferrin: A comparative study. *Biochim. Biophys. Acta* 257:314-323.
6. Al-Mashikhi, S.A., and Nakai, S. 1987. Isolation of bovine immunoglobulins and lactoferrin from whey proteins by gel filtration techniques. *J. Dairy Sci.* 70:2486-2492.
7. Alugupalli, K.R., Kalfas, S., Edwardsson, S., Forsgren, A., Arnold, R.R., and Naidu, A.S. 1994. Effect of lactoferrin on interaction of *Prevotella intermedia* with plasma and subepithelial matrix proteins. *Oral Microbiol. Immunol.* 9:174-179.
8. Alugupalli, K.R., and Kalfas, S. 1995. Inhibitory effect of lactoferrin on the adhesion of *Actinobacillus actinomycetemcomitans* and *Prevotella intermedia* to fibroblasts and epithelial cells. *APMIS* 103:154-160.
9. Alugupalli, K.R., Kalfas, S., Edwardsson, S., and Naidu, A.S. 1995. Lactoferrin interaction with *Actinobacillus actinomycetemcomitans*. *Oral Microbiol. Immunol.* 10:35-41.
10. Alugupalli, K.R., and Kalfas, S. 1997. Characterization of the lactoferrin-dependent inhibition of the adhesion of *Actinobacillus actinomycetemcomitans*, *Prevotella intermedia* and *Prevotella nigrescens* to fibroblasts and to a reconstituted basement membrane. *APMIS* 105:680-688.
11. Ambruso, D.R., Sasada, M., Nishiyama, H., Kubo, A., Komiyama, A., and Allen, R.H. 1984. Defective bactericidal activity and absence of specific granules in neutrophils from a patient with recurrent bacterial infections. *J. Clin. Immunol* . 4:23-30.
12. Amini, H.R., Ascencio, F., Ruiz-Bustos, E., Romero, M.J., and Wadstrom, T. 1996. Cryptic domains of a 60 kDa heat shock protein of *Helicobacter pylori* bound to bovine lactoferrin. *FEMS Immunol. Med. Microbiol.* 16: 247-55.
13. Andersson, B.F., Baker, H.M, Dodson, E.J., Norris, G.E., Rumball, S.V., Waters, J.M, and Baker E.N. 1987. Structure of human lactoferrin at 3.2-Å resolution. *Proc. Natl. Acad. Sci. USA.* 84:1769-1773.

14. Anderson, B.F., Baker, H.M., Norris, G.E., Rice, D.W., and Baker, E.N. 1989. Structure of human lactoferrin: crystallographic structure analysis and refinement at 2.8 A resolution. *J. Mol. Biol.* 209:711-734.
15. Appelmelk, B.J., An, Y.Q., Geerts, M., Thijs, B.G., de Boer, H.A., MacLaren, D.M., de Graaff, J., and Nuijens, J.H. 1994. Lactoferrin is a lipid A-binding protein. *Infect. Immun.* 62: 2628-2632.
16. Arnold, R.R., Cole, M.F., and McGhee, J.R. 1977. A bactericidal effect for human lactoferrin. *Science* 197:263-265.
17. Arnold, R.R., Brewer, M., and Gauthier, J.J. 1980. Bactericidal activity of human lactoferrin: sensitivity of a variety of microorganisms. *Infect. Immun.* 28:893-898.
18. Arnold, R.R., Russell, J.E., Champion, W.J., and Gauthier, J.J. 1981. Bactericidal activity of human lactoferrin: influence of physical conditions and metabolic state of the target microorganism. *Infect. Immun.* 32:655-660.
19. Ascencio, F., Ljungh, A., and Wadstrom, T. 1992. Lactoferrin binding properties of *Vibrio cholerae. Microbios.* 70:103-117.
20. Ashorn, R.G., Eskola, J., Tuohimaa, P.J., and Krohn, K.J. 1986. Effect of progesterone-inducible proteins human lactoferrin and avidin on lymphocyte proliferation. *Human Reproduct.* 1:149-151.
21. Aslam, M., and Hurley, W.L. 1997. Proteolysis of milk proteins during involution of the bovine mammary gland. *J. Dairy Sci.* 80:2004-2010.
22. Azari, P., and Baugh, R.F. 1967. A simple and rapid procedure for preparation of large quantities of pure ovotransferrin. *Arch. Biochem. Biophys.* 118:138-145.
23. Baggiolini, M., De Duve, C., Masson, P.L., and Heremans, J.F. 1970. Association of lactoferrin with specific granules in rabbit heterophil leukocytes. *J. Exp. Med.* 131:559-570.
24. Baker, E.N., Rumball, S.V., and Andersson, B.F. 1987. Transferrins: insight into structure and function from studies on lactoferrin. *Trends Biochem. Sci.* 12:350-353.
25. Baker, E.N., Anderson, B.F., Baker, H.M., Haridas, M., Jameson, G.B., Norris, G.E., Rumball, S.V., and Smith, C.A. 1991. Structure, function and flexibility of human lactoferrin. *Int. J. Biol. Macromol.* 13:122-129.
26. Balmer, S.E., Scott, P.H., and Wharton, B.A. 1989. Diet and feacal flora in the newborn: lactoferrin. *Arch Dis Child.* 64:1685-1690.
27. Barresi, G., and Tuccari, G. 1987. Iron-binding proteins in thyroid tumours: an immunocytochemical study. *Pathol. Res. Pract.* 182:344-351.
28. Barthe, C., Galabert, C., Guy-Crotte, O., and Figarella, C. 1989. Plasma and serum lactoferrin levels in cystic fibrosis. Relationship with the presence of cystic fibrosis protein. *Clin. Chim. Acta* 181:183-188.
29. Batish, V.K., Chander, H., Zumdegni, K.C., Bhatia, K.L., and Singh, R.S. 1988. Antibacterial effect of lactoferrin against some common food-borne pathogenic organisms. *Austral. J. Dairy Technol.* 5:16-18.
30. Baynes, R.D., Bezwoda, W.R., and Mansoor. N. 1988. Neutrophil lactoferrin content in viral infections. *Am. J. Clin. Pathol.* 89:225-228.
31. Bellamy, W., Takase, M., Yamauchi, K., Wakabayashi, H., Kawase, K., and Tomita, M. 1992. Identification of the bactericidal domain of lactoferrin. *Biochim. Biophys. Acta* 1121:130-136.
32. Bellamy, W., Wakabayashi, H., Takase, M., Kawase, K., Shimamura, S., and Tomita, M. 1993. Killing of *Candida albicans* by lactoferricin B, a potent antimicrobial peptide derived from the N-terminal region of bovine lactoferrin. *Med. Microbiol. Immunol. (Berl)* 182:97-105.
33. Bennett, R.M., and Kokocinski, T. 1978. Lactoferrin content of peripheral blood cells. *Br. J. Haematol.* 39:509-521.
34. Bennett, R.M., and Mohla, C. 1976. A solid phase radioimmunoassay for the measurement of lactoferrin in human plasma: variation with age, sex and disease. *J. Lab. Clin. Med.* 88:156-166.
35. Bennett, R.M., and Skosey, J.L. 1977. Lactoferrin and lysozyme levels in synovial fluid: differential indices of articular inflammation and degradation. *Arthrit. Rheumat.* 20:84-90.
36. Bennett, R.M., and Kokocinski, T. 1979. Lactoferrin turnover in man. *Clin. Sci.* 57:453-460.
37. Bennett, R.M., Bagby, G.C., and Davis, J. 1981. Calcium-dependent polymerization of lactoferrin. *Biochem. Biophys. Res. Commun.* 101:88-95.
38. Bennett, R.M., and Davis, J. 1982. Lactoferrin interacts with deoxyribonucleic acid: a preferential reactivity with double-stranded DNA and dissociation of DNA-anti-DNA complexes. *J. Lab. Clin. Med.* 99:127-38.
39. Bennett, R.M., Davis, J., Campbell, S., and Portnoff, S. 1983. Lactoferrin binds to cell membrane DNA. Association of surface DNA with an enriched population of B cells and monocytes. *J. Clin. Invest.* 71:611-618.
40. Bennett, R.M., Merritt, M.M., and Gabor, G. 1986. Lactoferrin binds to neutrophilic membrane DNA. *Br. J. Haematol.* 63:105-117.

41. Bezwoda, W.R., and Mansoor, H. 1986. Isolation and characterization of lactoferrin separated from human whey by adsorption chromatography using Cibacron Blue F3G-A linked affinity adsorbent. *Clin. Chim. Acta* 157:89-95.
42. Bezwoda, W.R., and Mansoor, N. 1989. Lactoferrin from human breast milk and from neutrophil granulocytes. Comparative studies of isolation, quantitation, characterization and iron binding properties. *Biomed. Chromat.* 3:121-126.
43. Bhimani, R.S., Vendrov, Y., and Furmanski, P. 1999. Influence of lactoferrin feeding and injection against systemic staphylococcal infections in mice. *J. Appl. Microbiol.* 86:135-144.
44. Birgens, H.S., and Kristensen, L. O. 1990. Impaired receptor binding and decrease in isoelectric point of lactoferrin after interaction with human monocytes. *Eur. J. Haemat* 45:31-35.
45. Bishop, J.G., Schanbacher, F.L., Ferguson, L.C., and Smith, K.L. 1976. *In vitro* growth inhibition of mastitis-causing coliform bacteria by bovine apo-lactoferrin and reversal of inhibition by citrate and high concentrations of apo lactoferrin. *Infect. Immun.* 14:911-918.
46. Biswas, G.B., and Sparling, P.F. 1995. Characterization of *lbpA*, the structural gene for a lactoferrin receptor in *Neisseria gonorrhoeae. Infect. Immun.* 63:2958-2967.
47. Blackberg, L., and Hernell, O. 1980. Isolation of lactoferrin from human whey by a single chromatography step. *FEBS Letts.* 109:180-184.
48. Blake, D.R., Hall, N.D., Bacon, P.A. 1981. The importance of iron in rheumatoid disease. *Lancet* 21:1142-1143.
49. Blake, D.R., Gallagher, P.J., and Potter, A.R. 1984. The effect of synovial iron on the progression of rheumatoid disease. *Arthritis Rheum.* 27:495-501.
50. Blanc, B., and Isliker, H.C. 1961a. Isolement et caracterisation de la proteine rouge siderophile dulait maternel: la lactoferrine. *Helv. Physiol. Pharmacol. Acta* 19:C-13-14.
51. Blanc, B., and Isliker, H.C. 1961b. Isolement et caracterisation de la proteine rouge siderophile dulait maternel: la lactoferrine. *Bull. Soc. Chim. Biol (Paris)* 43:929-943.
52. Bortner, C.A., Miller, R.D., and Arnold, R.R. 1986. Bactericidal effect of lactoferrin on *Legionella pneumophila. Infect. Immun.* 51:373-377.
53. Bortner, C.A., Arnold, R.R., and Miller, R.D. 1989. Bactericidal effect of lactoferrin on *Legionella pneumophila*: effect of the physiological state of the organism. *Can. J. Microbiol.* 35:1048-1051.
54. Boxer, L.A., Bjorksten, B., Bjork, J., Yang, H.H., Alen, J.M., and Baehner, R.L. 1982a. Neutropenia induced by systemic infusion of lactoferrin. *J. Lab. Clin. Med.* 99:866-872.
55. Boxer, L.A., Haak, R.A., Yang, H.H., Wolach, J.B., Whitcomb, J.A., Butterick, C.J., and Baehner, R.L. 1982b. Membrane-bound lactoferrin alters the surface properties of polymorphonuclear leukocytes. *J. Clin. Invest.* 70:1049-1057.
56. Breau, W.C., and Oliver, S.P. 1986. Growth inhibition of environmental mastitis pathogens during physiologic transitions of the bovine mammary gland. *Am. J. Vet. Res.* 47:218-222.
57. Brines, R.D., and Brock, J.H. 1983. The effect of trypsin and chymotrypsin on the *in vitro* antimicrobial and iron-binding properties of lactoferrin in human milk and bovine colostrum. Unusual resistance of human apolactoferrin to proteolytic digestion. *Biochim. Biophys. Acta* 759:229-235.
58. Britigan, B.E., Hayek, M.B., Doebbeling, B.N., and Fick, R.B. 1993. Transferrin and lactoferrin undergo proteolytic cleavage in the *Pseudomonas aeruginosa*-infected lungs of patients with cystic fibrosis. *Infect. Immun.* 61:5049-5055.
59. Britigan, B.E., Serody, J.S., and Cohen, M.S. 1994. The role of lactoferrin as an anti-inflammatory molecule. *Adv. Exp. Med.* Biol. 357:143-156.
60. Brock, J.H., Arzabe, F., Lampreave, F, and Pineiro, A. 1976. The effect of trypsin on bovine transferrin and lactoferrin. *Biochim. Biophys. Acta* 446:214-225.
61. Brock, J.H., Pineiro, A., and Lampreave, F. 1978. The effect of trypsin and chymotrypsin on the antibacterial activity of complement, antibodies, and lactoferrin and transferrin in bovine colostrum. *Ann. Rech. Vet.* 9:287-294.
62. Brock, J.H., Pickering, M.G., McDowall, M.C., and Deacon, A.G. 1983. Role of antibody and enterobactin in controlling growth of *Escherichia coli* in human milk and acquisition of lactoferrin- and transferrin-bound iron by *Escherichia coli. Infect. Immun.* 40:453-459.
63. Brodie, A.M., Ainscough, E.W., Baker, E.N., Baker, H.M., Shongwe, M.S., and Smith, C.A. 1994. Synergism and substitution in the lactoferrins. *Adv. Exp. Med. Biol.* 357:33-44.
64. Broer, K.H., Dauber, U., Herrmann, W.P., and Hirshhauser, C. 1977. The presence of lactoferrin (SCA) on the human sperm head during *in vitro* penetration through cervical mucus. *IRCS J. Med. Sci.* 5:362.
65. Broxmeyer, H.E., Smithyman, A., Eger, R.R., Meyers, P.A., and de Sousa, M. 1978. Identification of lactoferrin as the granulocyte-derived inhibitor of colony-stimulating activity production. *J. Exp. Med.* 148:1052-1067

66. Broxmeyer, H.E., and Platzer, E. 1984. Lactoferrin acts on I-A and I-E/C antigen subpopulations of mouse peritoneal macrophages in the absence of T lymphocytes and other cell types to inhibit production of granulocyte-monocyte colony stimulatory factors *in vitro*. *J. Immunol.* 133:306-314.
67. Broxmeyer, H.E., Williams, D.E., Hangoc, G., Cooper, S., Gentile, P., Shen, R.N., Ralph, P., Gillis, S., and Bicknell, D.C. 1987. The opposing action in *vivo* on murine myelopoiesis of purified preparations of lactoferrin and the colony stimulating factors. *Blood Cells* 13:31-48.
68. Bukhhardt, H., and. Schwingel, M. 1986. Oxygen radicals as effectors of cartilage destruction. *Arthritis Rheum.* 30:57-63.
69. Bullen, J.J. 1975. Iron-binding proteins in milk and resistance to *Escherichia coli* infection in infants. *Postgrad. Med. J.* 51:67-70.
70. Bullen, J.J. 1976. Iron-binding proteins and other factors in milk responsible for resistance to *Escherichia coli. Ciba Found. Symp.* 42:149-169.
71. Bullen, J.J. 1981. The significance of iron in infection. *Rev. Infect. Dis.* 3:1127-1138.
72. Burrin, D.G., Wang, H., Heath, J., and Dudley, M.A. 1996. Orally administered lactoferrin increases hepatic protein synthesis in formula-fed newborn pigs. *Pediatr. Res.* 40(1):72-76.
73. Butler, J.E. 1973. The occurrence of immunoglobulin fragments, two types of lactoferrin and a lactoferrin-IgG2 complex in bovine colostral and milk whey. *Biochim. Biophys. Acta* 295:341-351.
74. Byatt, J.C., Schmuke, J.J., Comens, P.G., Johnson, D.A., and Collier, R.J. 1990. The effect of bovine lactoferrin on muscle growth *in-vivo* and *in-vitro*. *Biochem. Biophys. Res. Comm.* 173:548-553.
75. Byrd, T.F., and Horwitz, M.A. 1991. Lactoferrin inhibits or promotes *Legionella pneumophila* intracellular multiplication in nonactivated and interferon gamma-activated human monocytes depending upon its degree of iron saturation. Iron-lactoferrin and nonphysiologic iron chelates reverse monocyte activation against *Legionella pneumophila. J. Clin. Invest.* 88:1103-1112.
76. Campagnari, A.A., Shanks, K.L., and Dyer, D.W. 1994. Growth of *Moraxella catarrhalis* with human transferrin and lactoferrin: expression of iron-repressible proteins without siderophore production. *Infect. Immun.* 62:4909-4914.
77. Carlsson, A., Bjorck, L., and Persson, K. 1989. Lactoferrin and lysozyme in milk during acute mastitis and their inhibitory effect in Delvotest P. *J. Dairy. Sci.* 72:3166-3175.
78. Carnoy, C., Scharfman, A., van Brussel, E., Lamblin, G., Ramphal, R., and Roussel, P. 1994. *Pseudomonas aeruginosa* outer membrane adhesins for human respiratory mucus glycoproteins. *Infect. Immun.* 62:1896-1900.
79. Caselitz, J., Jaup, T., and Seifer, G. 1981. Lactoferrin and lysozyme in carcinomas of the parotid gland. *Virchows Arch. A.* 394:61-73.
80. Castellino, F.J., Fish, W.W., and Mann, K.G. 1970. Structural studies on bovine lactoferrin. *J. Biol. Chem.* 245:4269-4275.
81. Charpin, C., Lachard, A., and Pourreau-Schneider, N. 1985. Localization of lactoferrin and nonspecific cross-reacting antigen in human breast carcinomas. *Cancer* 55:2612-2617.
82. Chierici, R., Sawatzki, G., Tamisari, L., Volpato, S., and Vigi, V. 1992. Supplementation of an adapted formula with bovine lactoferrin. 2. Effects on serum iron, ferritin and zinc levels. *Acta Paediatr.* 81:475-479.
83. Chimura, T., Hirayama, T., Nakahara, M. 1993. *In vitro* antimicrobial activities of lactoferrin, its concomitant use with cefpodoxime proxetil and clinical effect of cefpodoxime proxetil. *Jpn. J. Antibiot.* 46:482-485.
84. Cohen, M.S., Mao, J., Rasmussen, G.T., Serody, J.S., and Britigan, B.E. 1992. Interaction of lactoferrin and lipopolysaccharide (LPS): effects on the antioxidant property of lactoferrin and the ability of LPS to prime human neutrophils for enhanced superoxide formation. *J. Infect. Dis.* 166:1375-1378.
85. Courtoy, P.J., Moguilevsky, N., Retegui, L.A., Castracane, C.E., and Masson, P.L. 1984. Uptake of lactoferrin by the liver. II. Endocytosis by sinusoidal cells. *Lab. Invest.* 50:329-334.
86. Cox, T.M., Mazurier, J., Spik, G., Montreuil, J., and Peters, T.J. 1979. Iron-binding proteins and influx of iron across the duodenal brush border. Evidence for specific lactotransferrin receptors in the human small intestine. *Biochim. Biophys. Acta* 588:120-128.
87. Cravito, A., Tello, A., Villafan, H., Ruiz, J., del Vedovo, S., and Neeser, J.R. 1991. Inhibition of localized adhesion of enteropathogenic *Escherichia coli* to Hep-2 cells by immunoglobulin and oligosaccharide fractions of human colostrum and breast milk. *J. Infect. Dis.* 163:1247-1255.
88. Crosa, J.H. 1989. Genetics and molecular biology of siderophore-mediated iron transport in bacteria. *Microbiol. Rev.* 53:517-530.
89. Crouch, S.P.M., Slater, K.J., and Fletcher, J. 1992. Regulation of cytokine release from mononuclear cells by the iron-binding protein lactoferrin. *Blood* 80:235-240.

90. Culliere, M.L., Montagne, P., Mole, C., Bene, M.C., and Faure, G. 1997. Microparticle-enhanced nephlometric immunoassay of lactoferrin in human milk. *J. Clin. Lab. Anal.* 11:239-243.
91. Custer, M.C., and Hansen, J.N. 1983. Lactoferrin and transferrin fragments react with nitrite to form an inhibitor of *Bacillus cereus* spore outgrowth. *Appl. Environ. Microbiol.* 45:942-949.
92. Dalmastri, C., Valenti, P., Visca, P., Vittorioso, P., and Orsi, N. 1988. Enhanced antimicrobial activity of lactoferrin by binding to the bacterial surface. *Microbiologica* 11: 225-230.
93. Davidson, L.A., and Lönnerdal, B. 1986. Isolation and characterization of rhesus monkey lactoferrin. *Pediatr. Res.* 20:197-201.
94. Davidson, L., and Lönnerdal, B. 1988. Specific binding of lactoferrin to brush border membrane: Ontogeny and effect of glycan chain. *Am. J. Physiol.* 257:930-934.
95. Davidson, L.A., Litov, R.E., and Lönnerdal, B. 1990. Iron retention from lactoferrin-supplemented formulas in infant rhesus monkeys. *Pediatr. Res.* 27:176-180.
96. Davidson, L., Kastenmayer, P., Yuen, M., Lönnerdal, B., and Hurrell, R.F. 1994. Influence of lactoferrin on iron absorption from human milk in infants. *Pediatr. Res.* 35:117-124.
97. de Lillo, A., Teanpaisan, R., Fierro, J.F., and Douglas, C.W. 1996. Binding and degradation of lactoferrin by *Porphyromonas gingivalis, Prevotella intermedia* and *Prevotella nigrescens*. *FEMS Immunol. Med. Microbiol.* 14:135-143.
98. de Lillo, A., Quiros, L.M., and Fierro, J.F. 1997. Relationship between antibacterial activity and cell surface binding of lactoferrin in species of genus *Micrococcus*. *FEMS Microbiol. Lett.*150:89-94.
99. de Vet, B.J., and van Gool, J. 1974. Lactoferrin and iron absorption in the small intestine. *Acta Med. Scand.* 196:393-402.
100. de Vet, B.J., and van Vugt, H. 1971. Lactoferrin, in particular the relation with iron absorption. *Ned. Tijdschr. Geneeskd.* 115:961-967.
101. Debbabi, H., Dubarry, M., Rautureau, M., and Tome, D. 1998. Bovine lactoferrin induces both mucosal and systemic immune response in mice. *J. Dairy Res.* 65:283-293.
102. Denisova, I.I., Gennadeva, T.I., Zhdanova, T.V., Khristova, M., and Khazenson, L.B. 1996. The action of lactoperoxidase, lactoferrin and lactoglobulin on *Shigella sonnei* in an *in vivo* experiment. *Zh. Mikrobiol. Epidemiol. Immunobiol.* 1:66-69.
103. Dhaenens, L., Szczebara, F., and Husson, M.O. 1997. Identification, characterization, and immunogenicity of the lactoferrin-binding protein from *Helicobacter pylori*. *Infect. Immun.* 65:514-518.
104. Dial, E.J., Hall, L.R., Serna, H., Romero, J.J., Fox, J.G., and Lichtenberger, L.M. 1998. Antibiotic properties of bovine lactoferrin on *Helicobacter pylori*. *Dig. Dis. Sci.* 43:2750-2756.
105. Dinarello, C.A. 1984. Interleukin-1. *Rev. Infect. Dis.* 6:51-95.
106. Dionysius, D.A., and Milne, J.M. 1997. Antibacterial peptides of bovine lactoferrin: purification and characterization. *J. Dairy Sci.* 80:667-674.
107. Dionysius, D.A., Grieve, P.A., and Milne, J.M. 1993. Forms of lactoferrin: their antibacterial effect on enterotoxigenic *Escherichia coli*. *J. Dairy Sci.* 76:2597-2600.
108. Dite, P., Bartova, M., and Skaunic, V. 1989. Assessment of lactoferrin in the diagnosis of pancreatic disease. *Vnitrni. Lek.* 35:132-136.
109. Dolby, J.M., Stephens, S., and Honour, P. 1977. Bacteriostasis of *Escherichia coli* by milk. II. Effect of bicarbonate and transferrin on the activity of infant feeds. *J. Hyg. (Lond).* 78:235-242.
110. Dorland, L., Haverkamp, J., Vliegenthart, J.F.G., Spik, G., Fournet, B., and Montreuil, J. 1979. Investigation by 300-MHz 1H-nuclear magnetic resonance spectroscopy and methylation analysis of the single glycan chain of chicken ovotransferrin. *Eur. J. Biochem.* 100:569-574.
111. Elass-Rochard, E., Roseanu, A., Legrand, D., Trif, M., Salmon, V, Motas, C., Montreuil, J., and Spik G. 1995. Lactoferrin-lipopolysaccharide interaction: involvement of the 28-34 loop region of human lactoferrin in the high-affinity binding to *Escherichia coli* 055B5 lipopolysaccharide. *Biochem. J.* 312:839-45
112. Ellison, R.T., Giehl, T.J., and LaForce, F.M. 1988. Damage of the outer membrane of enteric gram-negative bacteria by lactoferrin and transferrin. *Infect. Immun.* 56:2774-2781.
113. Ellison, R.T., LaForce, F.M., Giehl, T.J., Boose, D.S., and Dunn, B.E. 1990. Lactoferrin and transferrin damage of the gram-negative outer membrane is modulated by Ca^{2+} and Mg^{2+}. *J. Gen. Microbiol.* 136:1437-1446.
114. Ellison, R.T., and Giehl, T.J. 1991. Killing of gram-negative bacteria by lactoferrin and lysozyme. *J. Clin. Invest.* 88:1080-1091.
115. Erdei, J., Forsgren, A., and Naidu, A.S. 1994. Lactoferrin binds to porins OmpF and OmpC in *Escherichia coli*. *Infect. Immun.* 62:1236-1240.
116. Faber, H.R., Bland, T., Day, C.L., Norris, G.E., Tweedie, J.W., and Baker, E.N. 1996. Altered domain closure and iron binding in transferrins: the crystal structure of the *Asp60Ser* mutant of the amino- terminal half-molecule of human lactoferrin. *J. Mol. Biol.* 256:352-363.

117. Fairweather-Tait, S.J., Balmer, S.E., Scott, P.H., and Minski, M.J. 1987. Lactoferrin and iron absorption in newborn infants. *Pediatr. Res.* 22:651-654.
118. Feng, M., van der Does, L., and Bantjes, A. 1995. Preparation of apolactoferrin with a very low iron saturation. *J. Dairy Sci.* 78:2352-2357.
119. Foster, L.A., and Dyer, D.W. 1993. A siderophore production mutant of *Bordetella bronchiseptica* cannot use lactoferrin as an iron source. *Infect. Immun.* 61:2698-2702.
120. Fransson, G.B., Keen, C.L., and Lonnerdal, B. 1983. Supplementation of milk with iron bound to lactoferrin using weanling mice: Effects on hematology and tissue iron. *J. Pediatr. Gastroenterol. Nutr.* 2:693-700.
121. Friedman, S.A., Mandel, I.D., and Herrera, M.S. 1983. Lysozyme and lactoferrin quantitation in the crevicular fluid. *J. Periodontol.* 54:347-350.
122. Fujihara, T., and Hayashi, K. 1995. Lactoferrin inhibits herpes simplex virus type-1 (HSV-1) infection to mouse cornea. *Arch. Virol.* 140:1469-72
123. Furmanski, P., Li, Z-P., Fortuna, M.B., Swamy, C.V.B., and Das, M.R. 1989. Multiple molecular forms of human lactoferrin. Identification of a class of lactoferrins that possess ribonuclease activity and lack iron-binding capacity. *J. Exp. Med.* 170:415-429.
124. Gado, I., Erdei, J., Laszlo, V.G., Paszti, J., Czirok, E., Kontrohr, T., Toth, I., Forsgren, A., and Naidu, A.S. 1991. Correlation between human lactoferrin binding and colicin susceptibility in *Escherichia coli*. *Antimicrob. Agents Chemother.* 35:2538-2543.
125. Gahr, M., Schulze, M., Schefeczyk, D., Speer, C.P., and Peters, J.H. 1987. Diminished release of lactoferrin from polymorphonuclear leukocytes in human neonates. *Acta Hematol.* 77:90-95.
126. Garre, C., Bianchi-Scarra, G., Sirito, M., Musso, M., and Ravazzolo, R. 1992. Lactoferrin binding sites and nuclear localization in K562(S) cells. *J. Cell. Physiol.* 153:477-482.
127. Gatter, K.C., Brown, G., Trowbridge, I.S., Woolston, R.E., and Mason, D.Y. 1983. Transferrin receptors in human tissues: their distribution and possible clinical relevance. *J. Clin. Pathol.*36:539-545.
128. Gillin, F.D., Reiner, D.S., and Wang, C.S. 1983. Killing of *Giardia lamblia* trophozoites by normal human milk. *J. Cell Biochem.* 23:47-56.
129. Giugliano, L.G., Ribeiro, S.T., Vainstein, M.H., and Ulhoa, C.J. 1995. Free secretory component and lactoferrin of human milk inhibit the adhesion of enterotoxigenic *Escherichia coli*. *J. Med. Microbiol.* 42:3-9.
130. Godo, Z.I., Magyar, E., Andirko, I., and Rozgonyi, F. 1997. Compact growth of *Staphylococcus haemolyticus* in soft agar is not due to hydrophobic interaction between the cocci. *Acta Microbiol. Immunol. Hung.* 44:343-349.
131. Gorden, W.G., Groves, M.L., and Basch, J.J. 1963. Bovine milk 'red protein': amino acid composition and comparison with blood transferrin. *Biochemistry* 2:817-820.
132. Griffiths, E., and Humphreys, J. 1977. Bacteriostatic effect of human milk and bovine colostrum on *Escherichia coli*: importance of bicarbonate. *Infect. Immun.* 15:396-401.
133. Groenink, J., Walgreen-Weterings, E., van't Hof, W., Veerman, E.C., and Nieuw Amerongen, A.V. 1999. Cationic amphipathic peptides, derived from bovine and human lactoferrins, with antimicrobial activity against oral pathogens. *FEMS Microbiol. Lett.* 179:217-222.
134. Grover, M., Giouzeppos, O., Schnagl, R.D., and May, J.T. 1997. Effect of human milk prostaglandins and lactoferrin on respiratory syncytial virus and rotavirus. *Acta. Paediatr.* 86:315-316.
135. Groves, M. 1960. The isolation of a red protein from milk. *J. Am. Chem. Soc.* 82:3345-3350.
136. Gruttner, R., Schafer, K.H., and Schroter, W. 1960. Zur reindarstellung eines eisenbindenden proteins in der frauenmilch. *Klinische Wochenschrift* 38:1162-1165.
137. Gutteberg, T.J., Haneberg, B., and Jørgensen, T. 1984. The latency of serum acute phase proteins in meningococcal septicemia, with special emphasis on lactoferrin. *Clin. Chim. Acta* 136:173-178.
138. Gutteberg, T.J., Røkke, O., Andersen, O., and Jørgensen, T. 1988. Lactoferrin as an indicator of septicemia and endotoxaemia in pigs. *Scand. J. Infect. Dis.* 20:659-666.
139. Gutteberg, T.J., Røkke, O., Andersen, O., and Jørgensen, T. 1989. Early fall of circulating iron and rapid rise of lactoferrin in septicemia and endotoxemia: an early defence mechanism. *Scand. J. Infect. Dis.* 21: 709-715.
140. Gutteberg, T.J., Osterud, B., Volden, G., and Jørgensen, T. 1990a. The production of tumour necrosis factor, tissue thromboplastin, lactoferrin and cathepsin C during lipopolysaccharide stimulation in whole blood. *Scand. J. Clin. Lab. Invest.* 50: 421-427.
141. Gutteberg, T.J., Dalaker, K., and Vorland, L.H. 1990b. Early response in neonatal septicemia. The effect of *Escherichia coli*, *Streptococcus agalactiae* and tumor necrosis factor on the generation of lactoferrin. *APMIS* 98:1027-1032.

142. Hamilton, H.L., and Czuprynski, C.J. 1992. Effects of mycobactin J and lactoferrin supplementation of drinking water on the *in vivo* multiplication of *Mycobacterium paratuberculosis* in gnotobiotic mice. *Can J Vet Res*. 56:70-73.
143. Hammerschmidt, S., Bethe, G., Remane, P., and Chhatwal, G.S. 1999. Identification of pneumococcal surface protein A as a lactoferrin-binding protein of *Streptococcus pneumoniae*. *Infect. Immun.* 67:1683-1687.
144. Hangoc, G., Lu, L., Oliff, A., Gillis, S., Hu, W., Bicknell, D.C., Williams, D., and Broxmeyer, H.E. 1987. Modulation of Friend virus infectivity *in vivo* by administration of purified preparations of human lactoferrin and recombinant murine interleukin-3 to mice. *Leukemia* 1: 762-764.
145. Hansen, N.E., Malmquist, J., and Thorell, J. 1975. Plasma myeloperoxidase and lactoferrin measured by radioimmunoassay: relations to neutrophil kinetics. *Acta Med. Scand.* 198:437-443.
146. Harbitz, O., Jensen, A.O., and Smidsrod, O. 1984. Lysozyme and lactoferrin in sputum from patients with chronic obstructive lung disease. *Eur. J. Res. Dis.* 65:512-520.
147. Harmon, R.J., Schanbacher, F.L., Fergusson, L.C., and Smith, K.L. 1976. Changes in lactoferrin, immunoglobulin G, bovine serum albumin, and alpha-lactalbumin during acute experimental and natural coliform mastitis in cows. *Infect. Immun.* 13:533-542.
148. Harmsen, M.C., Swart, P.J., de Bethune, M.P., Pauwels, R., De Clercq, E., The, T.H., and Meijer D.K. 1995. Antiviral effects of plasma and milk proteins: lactoferrin shows potent activity against both human immunodeficiency virus and human cytomegalovirus replication *in vitro*. *J. Infect. Dis.* 172:380-388.
149. Harris, W.R. 1986. Thermodynamics of gallium complexation by human lactoferrin. *Biochemistry* 25:803-808.
150. Hasegawa, K., Motsuchi, W., Tanaka, S., and Dosako, S. 1994. Inhibition with lactoferrin of *in vitro* infection with human herpes virus. *Jpn. J. Med. Sci. Biol.* 47:73-85
151. Hashizume, S., Kuroda, K., and Murakami, H. 1983. Identification of lactoferrin as an essential growth factor for human lymphocytic cell lines in serum-free medium. *Biochim. Biophys. Acta* 763:377-382.
152. He, J., and Furmanski, P. 1995. Sequence specificity and transcriptional activation in the binding of lactoferrin to DNA. *Nature* 373: 721-724
153. Hekman, A. 1971. Association of lactoferrin with other proteins, as demonstrated by changes in electrophoretic mobility. *Biochim. Biophys. Acta* 251:380-387.
154. Hentges, D.J., Marsh, W.W., Petschow, B.W., Thal, W.R., and Carter, M.K. 1992. Influence of infant diets on the ecology of the intestinal tract of human flora-associated mice. *J Ped Gastroenterol. Nutr.* 14:146-152.
155. Hirai, Y., Kawakata, N., Satoh, K., Ikeda, Y., Hisayasu, S., Orimo, H., and Yoshino, Y. 1990. Concentrations of lactoferrin and iron in human milk at different stages of lactation. *J. Nutr. Sci. Vitaminol. (Tokyo)* 36:531-544.
156. Hoek, K.S., Milne, J.M., Grieve, P.A., Dionysius, D.A., and Smith, R. 1997. Antibacterial activity in bovine lactoferrin-derived peptides. *Antimicrob. Agents Chemother.* 41:54-59.
157. Holmgren, J., Svennerholm, A.M., and Ahren, C. 1981. Nonimmunoglobulin fraction of human milk inhibits bacterial adhesion (hemagglution) and enterotoxin binding of *Escherichia coli* and *Vibrio cholerae*. *Infect. Immun.* 33:136-141.
158. Honour, P., and Dolby, J.M. 1979. Bacteriostasis of *Escherichia coli* by milk. III. The activity and stability of early, transitional and mature human milk collected locally. *J. Hyg. (Lond.).* 83: 243-254.
159. Hu, W.L., Mazurier, J., Montreuil, J., and Spik, G. 1990. Isolation and partial characterization of a lactotransferrin receptor from mouse intestinal brush border. *Biochemistry* 29:535-541.
160. Husson, M.O., Legrand, D., Spik, G., and Leclerc, H. 1993. Iron acquisition by *Helicobacter pylori*: importance of human lactoferrin. *Infect. Immun.* 61:2694-2697.
161. Hutchens, T.W., Magnuson, J.S., and Yip, T.T. 1989. Rapid purification of porcine colostral whey lactoferrin by affinity chromatography on single-stranded DNA-agarose. Characterization, amino acid composition, and N-terminal aminoacid sequence. *Biochim. Biophys. Acta* 999:323-329.
162. Iigo, M., Kuhara, T., Ushida, Y., Sekine, K., Moore, M.A., and Tsuda, H. 1999. Inhibitory effects of bovine lactoferrin on colon carcinoma 26 lung metastasis in mice. *Clin. Exp. Metastasis* 17:35-40.
163. Illingworth, D.S., Walter, K.S., Griffiths, P.L., and Barclay, R. 1993. Siderophore production and iron-regulated envelope proteins of *Helicobacter pylori*. *Zentralbl. Bakteriol.* 280:113-119.
164. Isamida, T., Tanaka, T., Omata, Y., Yamauchi, K., Shimazaki, K., and Saito, A. 1998. Protective effect of lactoferricin against *Toxoplasma gondii* infection in mice. *J. Vet. Med. Sci.* 60:241-244.
165. Izhar, M., Nuchamowitz, Y., and Mirelman, D. 1982. Adherence of *Shigella flexneri* to guinea pig intestinal cells is mediated by a mucosal adhesin. *Infect. Immun.* 35:1110-1118.
166. Janssen, P.T., and van Bijsterveld, O.P. 1982. Immunochemical determination of human tear lysozyme (muramidase) in keratoconjunctivitis sicca. *Clin. Chim. Acta.* 121:251-260.

167. Janssen, P.T., and van Bijsterveld, O.P. 1983. A simple test for lacrimal gland function: a tear lactoferrin assay by radial immunodiffusion. *Grafes. Arch. Clin. Exp. Opthalmol.* 220:171-174.
168. Jarosik, G.P., Land, C.B., Duhon, P., Chandler, R., and Mercer, T. 1998. Acquisition of iron by *Gardnerella vaginalis*. *Infect. Immun.* 66:5041-5047.
169. Johanson, B. 1960. Isolation of an iron-containing red protein from human milk. *Acta Chem. Scand.* 14:510-512.
170. Jolles, J., Donda, A., Amiguet, P., and Jolles, P. 1984. Mare lactotransferrin: purification, analysis and N-terminal sequence determination. *FEBS Lett.* 176:185-188.
171. Kalfas, S., Andersson, M., Edwardsson, S., Forsgren, A., and Naidu, A.S. 1991. Human lactoferrin binding to *Porphyromonas gingivalis, Prevotella intermedia,* and *Prevotella melaninogenica. Oral Microbiol. Immunol.* 6:350-355.
172. Kalfas, S., Tigyi, Z., Wikstrom, M., and Naidu, A.S. 1992. Laminin binding to *Prevotella intermedia. Oral Microbiol. Immunol.* 7:235-239.
173. Kalmar, J.R., and Arnold, R.R. 1988. Killing of *Actinobacillus actinomycetemcomitans* by human lactoferrin. *Infect. Immun.* 56:2552-2557.
174. Kang, J.H., Lee, M.K., Kim, K.L., and Hahm, K.S. 1996. Structure-biological activity relationships of 11-residue highly basic peptide segment of bovine lactoferrin. *Int. J. Pept. Protein Res.* 48:357-363.
175. Kawakami, H., and Lönnerdal, B. 1991. Isolation and function of a receptor for human lactoferrin in human fetal intestinal brush-border membranes. *Am. J. Physiol.* 261:G841-846.
176. Kawasaki, Y., Isoda, H., Shinmoto, H., Tanimoto, M., Dosako, S., Idota, T., and Nakajima, I. 1993. Inhibition by kappa-casein glycomacropeptide and lactoferrin of influenza virus hemagglutination. *Biosci. Biotechnol. Biochem.* 57:1214-1215.
177. Keen, C.L., Fransson, G.B., and Lonnerdal, B. 1984. Supplementation of milk with iron bound to lactoferrin using weanling mice: II. Effects on tissue manganese, zinc and copper. *J. Ped. Gastroenterol. Nutr.* 3:256-261.
178. Kennedy, R.S., Konok, G.P., Bounous, G., Baruchel, S., and Lee, T.D.G. 1995. The use of a whey protein concentrate in the treatment of patients with metastatic carcinoma:A phase I-II clinical study. *Anticancer Res.* 15:2643-2650.
179. Khin, M.M., Ringner, M., Aleljung, P., Wadstrom, T., and Ho, B. 1996. Binding of human plasminogen and lactoferrin by *Helicobacter pylori* coccoid forms. *J. Med. Microbiol.* 45:433-439.
180. Kijlstra, A.S., Jeurissen, H.M., and Koning, K.M. 1983. Lactoferrin levels in normal human tears. *Br. J. Opthalmol.* 67:199-202.
181. Kinkade, J.M., Miller III, W.W.K., and Segars, F.M. 1976. Isolation and characterization of murine lactoferrin. *Biochim. Biophys. Acta* 446:407-418.
182. Kioke, D., and Makino I. 1993. Protective effect of lactoferrin on caerulein-induced acute pancreatitis in rats. *Digestion.* 54:84-90.
183. Kirkpatrick, C.H., Green, I., Rich, R.R., and Schade, A.L. 1971. Inhibition of growth of *Candida albicans* by iron-unsaturated lactoferrin: relation to host-defense mechanisms in chronic mucocutaneous candidiasis. *J. Infect. Dis.* 124:539-544.
184. Kishore, AR., Erdei, J., Naidu, S.S., Falsen, E., Forsgren, A., Naidu, A.S. 1991. Specific binding of lactoferrin to *Aeromonas hydrophila. FEMS Microbiol. Lett.* 67:115-119.
185. Klasing, K.C. 1984. Effect of inflammatory agents and interleukin 1 on iron and zinc metabolism. *Am. J. Physiol.* 247:R901-R904.
186. Klebanoff, S.J., and Waltersdorph, A.M. 1990. Prooxidant activity of transferrin and lactoferrin. *J. Exp. Med.* 172:1293-1303.
187. Klempner, M.S, Dinarello, C.A., and Gallin, J.I. 1978. Human leukocytic progen induces release of specific granule contents from human neutrophils. *J. Clin. Invest.* 61:1330-1336.
188. Koivuranta-Vaara, P., Banda, D., and Goldstein, M. 1987. Bacterial-lipopolysaccharide-induced release of lactoferrin from human polymorphonuclear leukocytes: role of monocyte-derived tumor necrosis factor alpha. *Infect. Immun.* 55: 2956-61
189. Kokriakov, V.N., Alishina, G.M., Slepenkov, S.V., Lakovleva, M.F., and Pigarevskii, V.E. 1988. O Stepeni strukturnoi gomologii laktokerrinov moloka i neitrofil'nykh granulositov. *Biokhimiia* 53:1837-1843.
190. Krysteva, M., Mazurier, J., Spik, G., and Montreuil, J. 1975. Comparative study on histidine modification by diethylpyrocarbonate in human serotransferrin and lactotransferrin. *FEBS Lett.* 56:337-340.
191. Kulics, J. and Kijlstra, A. 1987. The effect of lactoferrin on complement mediated modulation of immune complex size. *Immunol. Lett.* 14:349-353.
192. Kume, S., and Tanabe, S. 1994. Effect of twinning and supplemental iron-saturated lactoferrin on iron status of newborn calves. *J. Dairy Sci.* 77:3118-3123

193. Kume, S., and Tanabe, S. 1995. Effect of supplemental lactoferrin with ferrous iron on iron status of newborn calves. *J. Dairy Sci.* 79:459-464.
194. Lampreave, F., Pineiro, A., Brock, J.H., Castillo, H., Sanchez, L., and Calvo, M. 1990. Interaction of bovine lactoferrin with other proteins of milk whey. *Int. J. Biol. Macromol.* 12:2-5.
195. Lash, J.A., Coates, T.D., Lauze, J., Baehner, R.L., and Boxer, L.A. 1983. Plasma lactoferrin reflects granulocyte activation *in vivo*. *Blood* 61:885-888.
196. Lassiter, M.O., Newsome, A.L., Sams, L.D., and Arnold, R.R. 1987. Characterization of lactoferrin interaction with *Streptococcus mutans*. *J. Dent. Res.* 66:480-485.
197. Lassiter, M.O. 1990. Ethylenediaminetetraacetate (EDTA): Influence on the antimicrobial activity of human lactoferrin against *Escherichia coli* and *Streptococcus mutans*. Doctoral Thesis. Department of Pathology, School of Medicine. Atlanta, Georgia: Emory University.
198. Leblebicioglu, B., Lim, J.S., Cario, A.C., Beck, F.M., and Walters, J.D. 1996. pH changes observed in the inflamed gingival crevice modulate human polymorphonuclear leukocyte activation *in vitro*. *J. Periodontol.* 67:472-477.
199. Lee, W.J., Farmer, J.L., Hilty, M., and Kim, Y.B. 1998. The protective effects of lactoferrin feeding against endotoxin lethal shock in germfree piglets. *Infect. Immun.* 66:1421-1426.
200. Leffell, M.S., and. Spitznagel, J.K. 1975. Fate of human lactoferrin and myeloperoxidase in phagocytosing human neutrophils: effects of immunoglobulin G subclasses and immune complexes coated on latex beads. *Infect. Immun.* 12:813-820.
201. Li, Y.M., Tan, A.X., and Vlassara, H. 1995. Antibacterial activity of lysozyme and lactoferrin is inhibited by binding of advanced glycation-modified proteins to a conserved motif. *Nat. Med.* 1:1057-1061.
202. Lima, M.F., and Kierszenbaum, F. 1985. Lactoferrin effects on phagocytic cell function. I. Increased uptake and killing of an intracellular parasite by murine macrophages and human monocytes. *J. Immunol.* 134:4176-4183.
203. Lima, M.F., and Kierszenbaum, F. 1987. Lactoferrin effects of phagocytic cell function. II. The presence of iron is required for the lactoferrin molecule to stimulate intracellular killing by macrophages but not to enhance the uptake of particles and microorganisms. *J. Immunol.* 139:1647-1651.
204. Lima, M.F., Beltz, L.A., and Kierszenbaum, F. 1988. *Trypanosoma cruzi*: a specific surface marker for the amastigote form. *J. Protozool.* 35: 108-110.
205. Longhi,C., Conte, M.P., Seganti, L., Polidoro, M., Alfsen, A., and Valenti, P. 1993. Influence of lactoferrin on the entry process of *Escherichia coli* HB101 (pRI203) in HeLa cells. *Med. Microbiol. Immunol.* 182:25-35.
206. Lönnerdal, B. 1989. Trace element absorption in infants as a foundation to setting upper limits for trace elements in infant formulas. *J. Nutr.(Suppl.)* 119:1839-1844.
207. Loughlin, K.R., and Gittes, R.F., Partridge, D., and Stelos, P. 1987. The relationship of lactoferrin to the anemia of renal cell carcinoma. *Cancer* 59:566-571.
208. Lu, L., Hangoc, G., Oliff, A., Chen, L.T., Shen, R.N., and Broxmeyer, H.E. 1987. Protective influence of lactoferrin on mice infected with the polycythemia-inducing strain of Friend virus complex. *Cancer Res.* 47:4184-4188.
209. Lugtenberg, B.,and van Alphen, L. 1983. Molecular architecture and functioning of the outer membrane of *Escherichia coli* and other gram-negative bacteria. *Biochim. Biophys. Acta* 737:51-115.
210. Luqmani, Y.A., Campbell, T.A., Bennett, C., Coombes, R.C., and Paterson, I.M. 1991. Expression of lactoferrin in human stomach. *Int. J. Cancer* 49:684-687.
211. Machnicki, M., Zimecki, M., and Zagulski, T. 1993. Lactoferrin regulates the release of tumour necrosis factor alpha and interleukin 6 *in vivo*. *Int. J. Exp. Pathol.* 74: 433-439.
212. Malmquist, J., and Johansson, B.G. 1971. Interaction of lactoferrin with agar gels and with trypan blue. *Biochim. Biophys. Acta* 236:38-46.
213. Mann, D.M., Romm, E., and Migliorini, M. 1994. Delineation of the glycosaminoglycan binding site in the human inflammatory response protein lactoferrin. *J. Biol. Chem.* 269:23661-23667.
214. Marchetti, M., Longhi, C., Conte, M.P., Pisani, S., Valenti, P., and Seganti, L. 1996. Lactoferrin inhibits herpes simplex virus type 1 adsorption to Vero cells. *Antiviral. Res.* 29:221-231.
215. Mason, D.Y., and Taylor, C.R. 1978. Distribution of transferrin, ferritin, and lactoferrin in human tissues. *J. Clin. Pathol.* 31:316-327.
216. Masson, P.L, Carbonara, A.O., and Heremans, J.F. 1965a. Studies on the proteins of human saliva. *Biochim. Biophys. Acta* 107:485-500.
217. Masson, P.L., Heremans, J.F., and Prignot, J., 1965b. Immunohistochemical localization of the iron-binding protein lactoferrin in human bronchial glands. *Experimentia (Basel)* 21:604-610.

218. Masson, P.L, Carbonara, A.O., and Heremans, J.F. 1965c. Studies on the proteins of human bronchial secretions. *Biochim. Biophys. Acta* 111:466.
219. Masson, P.L., Heremans, J.F., Prignot, J.J., and Wauters, G. 1966. Immunohistochemical localization and bacteriostatic properties of an iron-binding protein from bronchial mucus. *Thorax* 21:538-544.
220. Masson, P.L., Heremans, J.F., and Schonne, E. 1969. Lactoferrin, an iron-binding protein in neutrophilic leukocytes. *J. Exp. Med.* 130:643-658.
221. Masson, P.L., and Heremans, J.F. 1968. Metal-combining properties of human lactoferrin (red milk protein). 1. The involvement of bicarbonate in the reaction. *Eur J Biochem*. 6: 579-584.
222. Masson, P.L., and Heremans, J.F. 1971. Lactoferrin in milk from different species. *Comp. Biochem. Physiol.* 39B:119-129.
223. Mattsby-Baltzer, I., Roseanu, A., Motas, C., Elverfors, J., Engberg, I., and Hanson, L.A. 1996. Lactoferrin or a fragment thereof inhibits the endotoxin-induced interleukin-6 response in human monocytic cells. *Pediatr. Res.* 40:257-262.
224. Mazurier, J., and Spik, G. 1980. Comparative study of the iron-binding properties of human transferrins. I. Complete and sequential iron saturation and desaturation of the lactotransferrin. *Biochim. Biophys. Acta* 629:399-407.
225. Mazurier, J., Léger, D., Tordera, V., Montreuil, J., and Spik, G. 1981. Comparative study of the iron-binding properties of transferrins: Differences in the involvement of histidine residues as revealed by carbethoxylation. *Eur. J. Biochem.* 119:537-543.
226. McCormick, J.A., Markey, G.M., and Morris, T.C.M. 1991. Lactoferrin-inducible monocyte cytotoxicity for K562 cells and decay of natural killer lymphocyte cytotoxicity. *Clin. Exp.Immunol.* 83:154-156.
227. McMillan, J.A., Oski, F.A., Lourie, G., Tomarelli, R.M., and Landaw, S.A. 1977. Iron absorption from human milk, simulated human milk, and proprietary formulas. *Pediatrics* 60:896-900.
228. Menozzi, F.D., Ganteiz, C., and Locht, C. 1991. Identification and purification of transferrin and lactoferrin-binding proteins of *Bordetella pertussis* and *Bordetella bronchoseptica. Infect. Immun.* 59:3982-3988.
229. Metz-Boutigue, M.H., Jolles, J., Mazurier, J., Schoentgen, F., Legrand, D., Spik, G., Montreuil, J., and Jolles, P. 1984. Human lactotransferrin: amino acid sequence and structural comparisons with other transferrins. *Eur. J. Biochem.* 145:659-676.
230. Mickelsen, P.A., Blackman, E., and Sparling, P.F. 1982. Ability of *Neisseria gonorrhoeae, Neisseria meningitidis* and commensal *Neisseria* species to obtain iron from lactoferrin. *Infect. Immun.* 35:915-920.
231. Miedzobrodzki, J., Naidu, A.S., Watts, J.L., Ciborowski, P., Palm, K., and Wadstrom, T. 1989. Effect of milk on fibronectin and collagen type-I binding to *Staphylcoccus aureus* and coagulase-negative staphylococci isolated from bovine mastitis. *J. Clin. Microbiol.* 27:540-544.
232. Miyauchi, H., Kaino, A., Shinoda, I., Fukuwatari, Y., and Hayasawa, H. 1997. Immunomodulatory effect of bovine lactoferrin pepsin hydrolysate on murine splenocytes and Peyer's patch cells. *J. Dairy Sci.* 80:2330-2339.
233. Miyazawa, K., Mantel, C., Lu, L., Morrison, D.C., and Broxmeyer, H.E. 1991. Lactoferrin-lipopolysaccharide interactions. Effect on lactoferrin binding to monocyte/macrophage-differentiated HL-60 cells. *J. Immunol.* 146:723-729.
234. Moguilevsky, N., Retegui, L.A., and Masson, P.L. 1985. Comparison of human lactoferrins from milk and neutrophilic leukocytes: relative molecular mass, isoelectric point, iron-binding properties and uptake by the liver. *Biochem. J.* 229:353-359.
235. Montreuil, J., Tonnelat, J., and Mullet, S. 1960. Preparation and properties of lactoferrin of human milk. *Biochim. Biophys. Acta* 45:413-421.
236. Moore, S.A., Anderson, B.F., Groom, C.R., Haridas, M., and Baker, E.N. 1997. Three-dimensional structure of diferric bovine lactoferrin at 2.8 A resolution. *J. Mol. Biol.* 274:222-236.
237. Muller, F., Holberg-Petersen, M., Rollag, H., Degre, M., Brandtzaeg, P., and Froland, S.S. 1992. Nonspecific oral immunity in individuals with HIV infection. *J. Acquir. Immune Defic. Syndr.* 5:46-51.
238. Naidu, A.S., Miedzobrodzki, J., Andersson, M., Nilsson, L.E., Forsgren, A., and Watts, J.L. 1990. Bovine lactoferrin binding to six species of coagulase-negative staphylococci isolated from bovine intramammary infections. *J. Clin. Microbiol.* 28:2312-2319.
239. Naidu, A.S., Miedzobrodzki, J., Musser, J.M., Rosdahl, V.T., Hedstrom, S.A., and Forsgren, A. 1991a. Human lactoferrin binding in clinical isolates of *Staphylococcus aureus*. *J. Med. Microbiol.* 34:323-328.
240. Naidu, S.S., Erdei, J., Czirok, E., Kalfas, S., Gado, I., Thoren, A., Forsgren, A., and Naidu, A.S. 1991b. Specific binding of lactoferrin to *Escherichia coli* isolated from human intestinal infections. *APMIS*. 99:1142-1150.
241. Naidu, A.S., Andersson, M., and Forsgren, A. 1992. Identification of a human lactoferrin-binding protein in *Staphylococcus aureus*. *J. Med. Microbiol.* 36:177-183.

242. Naidu, S.S., Svensson, U., Kishore, A.R., and Naidu, A.S. 1993. Relationship between antibacterial activity and porin binding of lactoferrin in *Escherichia coli* and *Salmonella typhimurium*. *Antimicrob. Agents Chemother.* 37:240-245.
243. Naidu, A.S., and Arnold, R.R. 1994. Lactoferrin interaction with salmonellae potentiates antibiotic susceptibility in vitro. *Diagn. Microbiol. Infect. Dis.* 20:69-75.
244. Naidu, A.S. and Arnold, R.R. 1997. Influence of lactoferrin on host-microbe interactions. In *Lactoferrin - Interactions and Biological Functions*, ed. T.W. Hutchens, and B. Lonnerdal, 259-275. Totowa, NJ: Humana Press.
245. Naidu, A.S., and Bidlack, W.R. 1998. Milk lactoferrin - Natural microbial blocking agent (MBA) for food safety. *Environ. Nutr. Interact.* 2:35-50.
246. Nichols, B.L., McKee, K.S., and Huebers, H.A. 1990. Iron is not required in the lactoferrin stimulation of thymidine incorporation into the DNA of rat crypt enterocytes. *Pediatr. Res.* 27:525-528.
247. Nicholson, H., Anderson, B.F., Bland, T., Shewry, S.C., Tweedie, J.W., and Baker, E.N. 1997. Mutagenesis of the histidine ligand in human lactoferrin: iron binding properties and crystal structure of the histidine-253 methionine mutant. *Biochemistry* 36:341-346.
248. Nikaido, H., and Vaara, M. 1985. Molecular basis of bacterial outer membrane permeability. *Microbiol. Rev.* 49:1-31.
249. Nikaido, H. 1989. Outer membrane barrier as a mechanism of antimicrobial resistance. *Antimicrob. Agents Chemother.* 33:1831-1836.
250. Nishiya, K. and Horwitz, P.A. 1982. Contrasting effects of lactoferrin on human lymphocyte and monocyte natural killer cell activity and antibody-dependent cell mediated cytotoxicity. *J. Immunol.* 129:2519-2523.
251. Nonnecke, B.J., and Smith, K.L. 1984a. Inhibition of mastitic bacteria by bovine milk apo-lactoferrin evaluated by *in vitro* microassay of bacterial growth. *J. Dairy Sci.* 67:606-613.
252. Nonnecke, B.J., and Smith, K.L. 1984b.Biochemical and antibacterial properties of bovine mammary secretion during mammary involution and at parturition. *J. Dairy Sci.* 67:2863-2872.
253. Norris, G.E., Andersson, B.F., Baker, E.N., Baker, H.M., Gartner, A.L., Ward, J., and Rumball, S.V. 1986. Preliminary crystallographic studies on bovine lactoferrin. *J. Mol. Biol.* 191:143-145.
254. Odell, E.W., Sarra, R., Foxworthy, M., Chapple, D.S., and Evans, R.W. 1996. Antibacterial activity of peptides homologous to a loop region in human lactoferrin. *FEBS Lett.* 382:175-178.
255. Okujo, N., Akiyama, T., Miyoshi, S., Shinoda, S., and Yamamoto, S. 1996. Involvement of vulnibactin and exocellular protease in utilization of transferrin- and lactoferrin-bound iron by *Vibrio vulnificus*. *Microbiol. Immunol.* 40:595-598.
256. Okutomi, T., Tanaka, T., Yui, S., Mikami, M., Yamazaki, M., Abe, S., and Yamaguchi, H. 1998. Anti-*Candida* activity of calprotectin in combination with neutrophils or lactoferrin. *Microbiol. Immunol.* 42:789-793.
257. Olofson, T., Olsson, I., Venge, P., and Elgefors, B. 1977. Serum myeloperoxidase and lactoferrin in neutropenia. *Scand. J. Haematol.* 18:73-80.
258. Olusanya, O., and Naidu, A.S. 1991. Influence of mucin on serum and connective tissue protein binding to *Staphylococcus aureus* isolated from nasal carriage and clinical sources. *Afr. J. Med. Sci.* 20:89-95.
259. Oram, J.D. and Reiter, B. 1968. Inhibition of bacteria by lactoferrin and other iron-chelating agents. *Biochim. Biophys. Acta* 170:351-365.
260. Osborne, T.B., and Campbell, G.F. 1900. The protein constituents of egg white. *J. Am. Chem. Soc.* 22:422-426.
261. Oseas, R., Yang, H.H., Baehner, R.L., and Boxer, L.A. 1981. Lactoferrin: a promoter of polymorphonuclear leukocyte adhesiveness. *Blood* 57:939-945.
262. Otto, B.R., Verweij-van Vught, A.M., and MacLaren, D.M. 1992. Transferrins and heme-compounds as iron sources for pathogenic bacteria. *Crit. Rev. Microbiol.* 18:217-233.
263. Palma, C., Serbousek, D., Torosantucci, A., Cassone, A., and Djeu, J.Y. 1992. Identification of a mannoprotein fraction from *Candida albicans* that enhances human polymorphonuclear leukocyte (PMNL) functions and stimulates lactoferrin in PMNL inhibition of candidal growth. *J. Infect. Dis.* 166:1103-1112.
264. Paulsson, M.A., Svensson, U., Kishore, A.R., and Naidu, A.S. 1993. Thermal behavior of bovine lactoferrin in water and its relation to bacterial interaction and antibacterial activity. *J. Dairy Sci.* 76:3711-3720.
265. Payne, K.D., Davidson, P.M., Oliver, S.P., and Christian, G.L. 1990. Influence of bovine lactoferrin on the growth of Listeria monocytogenes. *J. Food Prot.* 53:468-472.
266. Payne, K.D., Davidson, P.M., and Oliver, S.P. 1994. Comparison of EDTA and apo-lactoferrin with lysozyme on the growth of food-borne pathogenic and spoilage bacteria. *J. Food Prot.* 57:62-65.
267. Pelus, L.M., Broxmeyer, H.E., and Moore, M.A. 1981. Regulation of human myelopoiesis by prostaglandin E and lactoferrin. *Cell Tissue Kinet.* 14:515:526.

268. Pentecost, B.T., and Teng, C.T. 1987. Lactotransferrin is the major estrogen inducible protein of mouse uterine secretions. *J. Biol. Chem.* 262:10134-10139.
269. Perraudin, J.P., Prieels, J.P., and Leonis, J. 1974. Interaction between lysozyme and some lactoferrin complex in human milk. *Arch. Int. Physiol. Biochim.* 82: 1001.
270. Peterson, K.M., and Alderete, J.F. 1984. Iron uptake and increased intracellular enzyme activity follow host lactoferrin binding by *Trichomonas vaginalis* receptors. *J. Exp Med.* 160:398-410.
271. Pettersson, A., Maas, A., and Tommassen, J. 1994. Identification of the iroA gene product of *Neisseria meningitidis* as a lactoferrin receptor. *J. Bacteriol.* 176:1764-1766.
272. Querinjean, P., Masson, P.L., and Heremans, J.F. 1971. Molecular weight, single-chain structure and amino acid composition of human lactoferrin. *Eur. J. Biochem.* 20:420-425.
273. Rainard, P. 1986a. Bacteriostasis of *Escherichia coli* by bovine lactoferrin, transferrin and immunoglobulins (IgG1, IgG2, IgM) acting alone or in combination. *Vet. Microbiol.* 11:103-115.
274. Rainard, P. 1986b. Bacteriostatic activity of bovine milk lactoferrin against mastitic bacteria. *Vet. Microbiol.* 11:387-392.
275. Rainard, P. 1987. Bacteriostatic activity of bovine lactoferrin in mastitic milk. *Vet. Microbiol.* 13: 159-166.
276. Rainard, P. 1992. Binding of bovine lactoferrin to *Streptococcus agalactiae*. *FEMS Microbiol. Lett.* 77:235-239.
277. Rainard, P. 1993. Activation of the classical pathway of complement by binding of bovine lactoferrin to unencapsulated *Streptococcus agalactiae*. *Immunology* 79:648-652.
278. Raphael, G.D., Jeney, E.V., Baranuik, J.N., Kim, I., Meridith, S.D., and Kaliner, M.A. 1989. Pathophysiology of rhinitis: lactoferrin and lysozyme in nasal secretion. *J. Clin. Invest.* 84:1528-1535.
279. Regoeczi, E., Chindemi, P.A., Debanne, M.T., and Prieels, J.P. 1985. Lactoferrin catabolism in the rat liver. *Am.J. Physiol.* 248:G8-G14.
280. Regoeczi, E., Chindemi, P.A., and Hu, W.L. 1994. Transport of lactoferrin from blood to bile in the rat. *Hepatology* 19:1476-1482.
281. Reiter, B., and Oram, J.D. 1967. Bacterial inhibitors in milk and other biological fluids. *Nature* 216:328-330.
282. Reiter, B., Brock, J.H., and Steel, E.D. 1975. Inhibition of *Escherichia coli* by bovine colostrum and post-colostral milk. II. The bacteriostatic effect of lactoferrin on a serum susceptible and serum resistant strain of *E. coli*. *Immunology* 28:83-95.
283. Rejman, J.J., Hegarty, H.M., and Hurley, W.L. 1989. Purification and characterization of bovine lactoferrin from secretions of the involuting mammary gland: Identification of multiple molecular forms. *Comp. Biochem. Physiol.* 93B:929-934.
284. Roberts, A.K., Chierici, R., Hill, M.J., Volpata, S., and Vigi, V. 1992. Supplementation of an adapted formula with bovine lactoferrin: I. Effect on the infant fecal flora. *Acta Paediatr.* 81:119-124.
285. Rogers, H.J., and Synge, C. 1978. Bacteriostatic effect of human milk on *Escherichia coli*: the role of IgA. *Immunology* 34:19-28.
286. Rogers, T.B., Borresen, T., and Feeney, R.E. 1978. Chemical modification of the arginines in transferrins. *Biochemistry* 17:1105-1109.
287. Rosa, G., and Trugo, N.M. 1994. Iron uptake from lactoferrin by intestinal brush-border membrane vesicles of human neonates. *Braz. J. Med. Biol. Res.* 27:1527-1531.
288. Rumke, P., Visser, D., Kwa, H.G., and Hart, A.A.M.. 1971. Radioimmunoassay for lactoferrin in blood plasma of breast cancer patients, lactating, and normal women: prevention of false high levels caused by leakage from neutrophil leukocytes in vitro. *Folia Med. Netherland.* 14:156-168.
289. Rumke, P. 1974. The origin of immunoglobulins in semen. *Clin. Exp. Med.* 14:1068-1084.
290. Saito, H., Takase, M., Tamura, Y., Shimamura, S., and Tomita, M. 1994. Physicochemical and antibacterial properties of lactoferrin and its hydrolysate produced by heat treatment at acidic pH. *Adv. Exp. Med. Biol.* 357:219-226.
291. Salamah, A.A., and al-Obaidi, A.S. 1995a. *In vivo* and *in vitro* effects of lactoferrin on *Yersinia pseudotuberculosis*. *New Microbiol* 18:267-274.
292. Salamah, A.A., and al-Obaidi, A.S. 1995b. Effect of some physical and chemical factors on the bactericidal activity of human lactoferrin and transferrin against *Yersinia pseudotuberculosis*. *New Microbiol* 18:275-281.
293. Samaranayake, Y.H., Samaranayake, L.P., Wu, P.C., and So, M. 1997. The antifungal effect of lactoferrin and lysozyme on *Candida krusei* and *Candida albicans*. *APMIS* 105:875-883.
294. Sanchez, L., Ismail, M., Liew, F.Y., and Brock, J.H. 1996. Iron transport across Caco-2 cell monolayers. Effect of transferrin, lactoferrin and nitric oxide. *Biochim. Biophys. Acta* 1289:291-297.

295. Sato, N., Kurotaki, H., Ikeda, S., Daio, R., Nishinome, N., Mikami, T., and Matsumoto, T. 1999. Lactoferrin inhibits *Bacillus cereus* growth and heme analogs recover its growth. *Biol. Pharm. Bull.* 22:197-199.
296. Sato, R., Inanami, O., Tanaka, Y., Takase, M., and Naito, Y. 1996. Oral administration of bovine lactoferrin for treatment of intractable stomatitis in feline immunodeficiency virus (FIV)-positive and FIV-negative cats. *Am. J. Vet. Res.* 57:1443-1446.
297. Sawatzki, G., and Rich, I.N. 1989. Lactoferrin stimulates colony stimulating factor production *in vitro* and *in vivo*. *Blood Cells* 15: 371-385.
298. Schade, A.L., and Caroline, L. 1946. An iron-binding component in human blood plasma. *Science* 104:340.
299. Schafer, K.H. 1951. Elekrophoretische untersuchungen zum milcheiweissproblem. *Montssschrift fur Kindrheilkunde* 99:69-71.
300. Schlabach, M.R. and Bates, G.W. 1975. The synergistic binding of anions and Fe(III) by transferrin. Implications for the interlocking sites hypothesis. *J. Biol. Chem.* 250:2182-2188.
301. Schryver, A.B., and Morris, L.J. 1988. Identification and characterization of the transferrin receptor from *Neisseria meningitidis*. *Mol. Microbiol.* 2:281-288.
302. Schryver, A.B. 1989. Identification of the transferrin- and lactoferrin-binding proteins in *Haemophilus influenzae*. *J. Med. Microbiol.* 29:121-130.
303. Schryver, A.B., and Stojiljkovic, I. 1999. Iron acquisition systems in the pathogenic *Neisseria*. *Mol. Microbiol.* 32:1117-1123.
304. Schulz-Lell, G., Dorner, K., Oldigs, H.D., Sievers, E., and Schaub, J. 1991. Iron availability from an infant formula supplemented with bovine lactoferrin. *Acta Paediatr. Scand.* 80:155-158.
305. Sekine, K., Watanabe, E., Nakamura, J., Takasuka, N., Kim, D.J., Asamoto, M., Krutovskikh, V., Baba-Toriyama, H., Ota, T., Moore, M.A., Masuda, M., Sugimoto, H., Nishino, H., Kakizoe, T., and Tsuda, H. 1997a. Inhibition of azoxymethane-initiated colon tumor by bovine lactoferrin administration in F344 rats. *Jpn. J. Cancer Res.* 88:523-526.
306. Sekine, K., Ushida, Y., Kuhara, T., Iigo, M., Baba-Toriyama, H., Moore, M.A., Murakoshi, M., Satomi, Y., Nishino, H., Kakizoe, T., and Tsuda, H. 1997b. Inhibition of initiation and early stage development of aberrant crypt foci and enhanced natural killer activity in male rats administered bovine lactoferrin concomitantly with azoxymethane. *Cancer Lett.* 121:211-216.
307. Shau, H., Kim, A., Golub, S.H. 1992. Modulation of natural killer and lymphokine-activated killer cell cytotoxicity by lactoferrin. *J. Leukoc. Biol.* 51:343-349.
308. Shimazaki, K., and Nishio, N. 1991. Interacting properties of bovine lactoferrin with immobilized Cibacron blue F3GA in column chromatography. *J. Dairy Sci.* 74:404-408.
309. Shimazaki, K., Kawaguchi, A., Sato, T., Ueda, Y., Tommimura, T., and Shimamura, S. 1993. Analysis of human and bovine milk lactoferrins by rotafor and chromatofocusing. *Int. J. Biochem.* 25:1653-1658.
310. Shimazaki, K., Nam, M.S., Harakawa, S., Tanaka, T., Omata, Y., Saito, A., Kumura, H., Mikawa, K., Igarashi, I., and Suzuki, N. 1996. Monoclonal antibody against bovine lactoferricin and its epitopic site. *J. Vet. Med. Sci.* 58:1227-1229.
311. Shimizu, K., Matsuzawa, H., Okada, K., Tazume, S., Dosako, S., Kawasaki, Y., Hashimoto, K., and Koga, Y. 1996. Lactoferrin-mediated protection of the host from murine cytomegalovirus infection by a T-cell-dependent augmentation of natural killer cell activity. *Arch. Virol.* 141:1875-1889.
312. Shin, K., Yamauchi, K., Teraguchi, S., Hayasawa, H., Tomita, M., Otsuka, Y., Yamazaki, S. 1998. Antibacterial activity of bovine lactoferrin and its peptides against enterohaemorrhagic *Escherichia coli* O157:H7. *Lett. Appl. Microbiol.* 26:407-411.
313. Shinmoto, H., Kobori, M., Tsushida, T., and Shinohara, K. 1997. Competitive ELISA of bovine lactoferrin with bispecific monoclonal antibodies. *Biosci. Biotechnol. Biochem.* 61:1044-1046.
314. Smith, K.L., Conrad, H.R., and Porter, R.M. 1971. Lactoferrin and IgG immunoglobulins from involuted bovine mammary glands. *J Dairy Sci.* 54:1427-1435.
315. Sojar, H.T., Hamada, N. and Genco, R.J. 1998. Structures involved in the interaction of *Porphyromonas gingivalis* fimbriae and human lactoferrin. *FEBS Lett.* 422:205-208.
316. Sørensen, M., and Sørensen, S.P.L. 1939. The proteins in whey. *CR. Trav. Lab. Carlsberg*. 23:55-99.
317. Soukka, T., Lumikari, M., and Tenovuo, J. 1991. Combined inhibitory effect of lactoferrin and lactoperoxidase system on the viability of *Streptococcus mutans*, serotype c. *Scand. J. Dent. Res.* 99: 390-396.
318. Spik, G., Bayard, B., Fournet, B., Strecker, G., Bouquelet, S., and Montreuil, J. 1975. Studies on glycoconjugates. LXIV. Complete structure of two carbohydrate units of human serotransferrin. *FEBS Lett.* 50:296-299.
319. Spik, G., Strecker, G., Fournet, B., Bouquelet, S., Montreuil, J., Dorland, L., Van Halbeek, H., and Vliegenthart, J.F.G. 1982. Primary structure of glycans from human lactotransferrin. *Eur. J. Biochem.* 121:413-419.

320. Staggs, T.M., Greer, M.K., Baseman, J.B., Holt, S.C., and Tyron, V.V. 1994. Identification of lactoferrin-binding proteins from *Treponema pallidum* subspecies *pallidum* and *Treponema denticola*. *Mol. Microbiol.* 12:613-619.
321. Stuart, J., Norrell, S., and Harrington, J.P. 1984. Kinetic effect of human lactoferrin on the growth of *Escherichia coli* 0111. *Int. J. Biochem.* 16:1043-1047.
322. Superti, F., Ammendolia, M.G., Valenti, P., and Seganti, L. 1997. Antirotaviral activity of milk proteins: lactoferrin prevents rotavirus infection in the enterocyte-like cell line HT-29. *Med. Microbiol. Immunol. (Berl.)* 186: 83-
323. Swart, P.J., Kuipers, M.E., Smit, C., Pauwels, R., deBethune, M.P., de Clercq, E., Meijer, D.K., and Huisman, J.G. 1996. Antiviral effects of milk proteins: acylation results in polyanionic compounds with potent activity against human immunodeficiency virus types 1 and 2 *in vitro*. *AIDS Res. Hum. Retroviruses.* 12:769-775.
324. Szuchet-Derechin, S., and Johnson, P. 1965. The 'Albumin' fraction of bovine milk-II. The isolation and physicochemical properties of the red proteins. *Eur. Polymer J.* 1:283-291.
325. Tabak, L., Mandel, I.D., Herrera, M., and Baurmash, H. 1978. Changes in lactoferrin and other proteins in a case of chronic recurrent parotitis. *J. Oral Path.* 1:97-99.
326. Tachezy, J., Kulda, J., Bahnikova, I., Suchan, P., Razga, J., and Schrevel, J. 1996. *Tritrichomonas foetus:* iron acquisition from lactoferrin and transferrin. *Exp. Parasitol.* 83:216-228.
327. Tanaka, T., Omata, Y., Saito, A., Shimazaki, K., Igarashi, I., and Suzuki, N. 1996. Growth inhibitory effects of bovine lactoferrin to *Toxoplasma gondii* parasites in murine somatic cells. *J. Vet. Med. Sci.* 58: 61-65.
328. Tashima, K.T., Carroll, P.A., Rogers, M.B., and Calderwood, S.B. 1996. Relative importance of three iron-regulated outer membrane proteins for *in vivo* growth of *Vibrio cholerae*. *Infect. Immun.* 64:1756-1761.
329. Teng, C. T., Walker, M. P., Bhattacharyya, S. N., Klapper, D. G., DiAugustine, R. P., and McLachlan, J. A. 1986. Purification and properties of an oestrogen-stimulated mouse uterine glycoprotein (approx. 70 kDa). *Biochem. J.* 240:413-422.
330. Teraguchi, S., Shin, K., Ozawa, K., Nakamura, S., Fukuwatari, Y., Tsuyuki, S., Namihira, H., and Shimamura, S. 1995a. Bacteriostatic effect of orally administered bovine lactoferrin on proliferation of *Clostridium* species in the gut of mice fed bovine milk. *Appl. Environ. Microbiol.* 61:501-506.
331. Teraguchi, S., Shin, K., Ogata, T., Kingaku, M., Kaino, A., Miyauchi, H., Fukuwatari, Y., and Shimamura, S. 1995b. Orally administered bovine lactoferrin inhibits bacterial translocation in mice fed bovine milk. *Appl. Environ. Microbiol.* 61:4131-4134.
332. Teraguchi, S., Shin, K., Fukuwatari, Y., and Shimamura, S. 1996. Glycans of bovine lactoferrin function as receptors for the type 1 fimbrial lectin of *Escherichia coli*. *Infect. Immun.* 64:1075-1077.
333. Teuwissen, B., Masson, P.L., Osinski, P., and Heremans, J.F. 1972. Metal-combining properties of human lactoferrin. The possible involvement of tyrosyl residues in the binding sites. Spectrophotometric titration. *Eur. J. Biochem.* 31:239-245.
334. Thomas, A.S., and Fell, L.R. 1985. Effect of ACTH and oxytocin treatment on lactoferrin and citrate in cows' milk. *J. Dairy Res.* 52:379-389.
335. Tigyi, Z., Kishore, A.R., Maeland, J.A., Forsgren, A., and Naidu, A.S. 1992. Lactoferrin-binding proteins in *Shigella flexneri*. *Infect. Immun.* 60:2619-2626.
336. Todhunter, D.A., Smith, K.L., and Schoenberger, P.S. 1985. *In vitro* growth of mastitis-associated streptococci in bovine mammary secretions. *J. Dairy Sci.* 68:2337-2346.
337. Toma, C., Honma, Y., and Iwanaga, M. 1996. Effect of *Vibrio cholerae* non-O1 protease on lysozyme, lactoferrin and secretory immunoglobulin A. *FEMS Microbiol. Lett.* 135:143-147.
338. Tomita, M., Bellamy, W., Takase, M., Yamauchi, K., Wakabayashi, H., and Kawase, K. 1991. Potent antibacterial peptides generated by pepsin digestion of bovine lactoferrin. *J. Dairy Sci.* 74:4137-4142.
339. Tomita, M., Takase, M., Wakabayashi, H., and Bellamy, W. 1994. Antimicrobial peptides of lactoferrin. *Adv. Exp. Med. Biol.* 357: 209-218.
340. Tomita, S., Shirasaki, N., Hayashizaki, H., Matsuyama, J., Benno, Y., and Kiyosawa, I. 1998. Binding characteristics of bovine lactoferrin to the cell surface of *Clostridium* species and identification of the lactoferrin-binding protein. *Biosci. Biotechnol. Biochem.* 62:1476-1482.
341. Tourville, D.R., S.S. Ogra, J. Lippes, and T.B. Tomassi Jr. 1970. The human female reproductive tract:immunohistological localization of gammaA, gammaG, gammaM "secretory piece" and lactoferrin. *Am. J. Obstet. Gynecol.* 108:1102-
342. Trumpler, U., Straub, P.W., and Rosenmund, A. 1989. Antibacterial prophylaxis with lactoferrin in neutropenic patients. *Eur. J. Clin. Microbiol. Infect. Dis.* 8:310-313.
343. Tryon, V.V., and Baseman, J.B. 1987. The acquisition of human lactoferrin by *Mycoplasma pneumoniae*. *Microb. Pathog.* 3:437-443.

344. Tsuji, S., Y. Hirata, and K. Matsuoka. 1989. Two apparent molecular forms of bovine lactoferrin. *J. Dairy Sci.* 72:1130-1136.
345. Tuccari, G., and G. Barresi. 1985. Immunohistochemical demonstration of lactoferrin in follicular adenomas and thyroid carcinomas. *Virchows Arch. A.* 406:67-74.
346. Tuccari, G., Barresi, G., Arena, F., and Inferrera, C. 1989. Immunocytochemical detection of lactoferrin in human gastric carcinomas and adenomas. *Arch. Pathol. Lab. Med.*113:912-915.
347. Turchany, J.M., Aley, S.B., and Gillin, F.D. 1995. Giardicidal activity of lactoferrin and N-terminal peptides. *Infect. Immun.* 63:4550-4552.
348. van Berkel, P.H., Geerts, M.E., van Veen, H.A., Mericskay, M., de Boer, H.A., and Nuijens, J.H..1997.N-terminal stretch *Arg2, Arg3, Arg4* and *Arg5* of human lactoferrin is essential for binding to heparin, bacterial lipopolysaccharide, human lysozyme and DNA. *Biochem. J.* 15:145-451.
349. Van Halbeck, H., Dorland, L., Vliegenthar, J.F.G., Spil, G., Cheron, A., and Montreuil, J. 1981. Structure determination of two oligomannoside-type glycopeptides obtained from bovine lactotransferrin, by 500 MHz 1H-NMR spectroscopy. *Biochim. Biophys. Acta* 675:293-296.
350. Van Snick, J., Masson, P.L., and Heremans, J.F. 1974. The involvement of lactoferrin in the hyposideremia of acute inflammation. *J. Exp. Med.* 140:1068-1084.
351. Van Snick, J., and Masson, P.L. 1976. The binding of human lactoferrin to mouse peritoneal cells. *J. Exp. Med.* 144:1568-1580.
352. Visca, P., Berlutti, F., Vittorioso, P., Dalmastri, C., Thaller, M.C., and Valenti, P. 1989. Growth and adsorption of *Streptococcus mutans* 6715-13 to hydroxyapatite in the presence of lactoferrin. *Med. Microbiol. Immunol. (Berl)* 178:69-79.
353. Visca, P., Dalmastri, C., Verzili, D., Antonini, G., Chiancone, E., Valenti, P. 1990. Interaction of lactoferrin with *Escherichia coli* cells and correlation with antibacterial activity. *Med. Microbiol. Immunol. (Berl)* 179:323-333.
354. Wada, T., Aiba, Y., Shimizu, K., Takagi, A., Miwa, T., and Koga, Y. 1999. The therapeutic effect of bovine lactoferrin in the host infected with *Helicobacter pylori. Scand. J. Gastroenterol.* 34:238-243.
355. Wakabayashi, H., Abe, S., Okutomi, T., Tansho, S., Kawase, K., and Yamaguchi, H. 1996. Cooperative anti-Candida effects of lactoferrin or its peptides in combination with azole antifungal agents. *Microbiol. Immunol.* 40:821-825.
356. Wang, C.S., Chan, W.Y., and Kloer, U.H. 1984. Comparative studies on the chemical and immunochemical properties of human milk, human pancreatic juice and bovine milk lactoferrin. *Comp. Biochem. Physiol.* (B) 78:575-580.
357. Wang, D., Pabst, K.M., Aida, Y., and Pabst, M.J. 1995. Lipopolysaccharide-inactivating activity of neutrophils is due to lactoferrin. *J. Leukoc. Biol.* 57:865-874.
358. Watanabe, T., Nagura, H., Watanabe, K., and Brown, W.R. 1984. The binding of human milk lactoferrin to immunoglobulin A. *FEBS Lett.* 168:203-207.
359. Weinberg, E.D. 1975. Nutritional immunity. Host's attempt to withold iron from microbial invaders. *JAMA.* 231 (1): 39-41.
360. Wichmann, L., Vaalasti, A., Vaalasti, T., and Tuohimaa, P. 1989. Localization of lactoferrin in the male reproductive tract. *Int. J. Androl.* 12:179-186.
361. Williams, P.H., and Carbonetti, N.H. 1986. Iron, siderophores, and the pursuit of virulence: independence of the aerobactin and enterochelin iron uptake systems in *Escherichia coli. Infect. Immun.* 51:942-947.
362. Wilson, M.E., Vorhies, R.W., Andersen, K.A., and Britigan, B.E. 1994. Acquisition of iron from transferrin and lactoferrin by the protozoan *Leishmania chagasi. Infect. Immun.* 62:3262-3269.
363. Windle, J.J., Wiersema, A.K., Clark, J.R., and Feeney, R.E. 1963. *Biochemistry* 2:1341-1346.
364. Wu, H.F., Monroe, D.M., and Church, F.C. 1995. Characterization of the glycosaminoglycan binding region of lactoferrin. *Arch. Biochem. Biophys.* 317:85-92.
365. Yamauchi, K., Tomita, M., Giehl, T.J., and Ellison, R.T. 1993. Antibacterial activity of lactoferrin and a pepsin-derived lactoferrin peptide fragment. *Infect. Immun.* 61:719-728.
366. Yamauchi, K., Wakabayashi, H., Hashimoto, S., Teraguchi, S., Hayasawa, H., and Tomita, M. 1998. Effects of orally administered bovine lactoferrin on the immune system of healthy volunteers. *Adv. Exp. Med. Biol.* 443:261-265.
367. Yang, H., Kooi, C.D., and Sokol, P.A. 1993. Ability of *Pseudomonas pseudomallei* malleobactin to acquire transferrin-bound, lactoferrin-bound, and cell-derived iron. *Infect. Immun.* 61:656-662.
368. Yi, M., Kaneko, S., Yu, D.Y., and Murakami, S. 1997. Hepatitis C virus envelope proteins bind lactoferrin. *J. Virol.* 71:5997-6002.
369. Yoo, Y.C., Watanabe, S., Watanabe, R., Hata, K., Shimazaki, K., and Azuma, I. 1997. Bovine lactoferrin and lactoferricin, a peptide derived from bovine lactoferrin, inhibit tumor metastasis in mice. *Jpn. J. Cancer Res.* 88:184-190.

370. Yoshida, S. 1989. Preparation of lactoferrin by hydrophobic interaction chromatography from milk acid whey. *J. Dairy Sci.* 72:1446-1450.
371. Yoshida, S., and Ye-Xiuyun. 1991. Isolation of lactoperoxidase and lactoferrins from bovine milk acid whey by carboxymethyl cation exchange chromatography. *J. Dairy Sci.* 74:1439-1444.
372. Zagulski, T., Lipinski, P., Zagulska, A., Broniek, S., and Jarzabek, Z. 1989. Lactoferrin can protect mice against a lethal dose of *Escherichia coli* in experimental infection *in vivo*. *Br. J. Exp. Pathol.* 70:697-704.
373. Zhao, X.Y., and Hutchens, T.W. 1994. Proposed mechanisms for the involvement of lactoferrin in the hydrolysis of nucleic acids. *Adv. Exp. Med. Biol.* 357: 271-278.
374. Ziere, G.J., van Dijk, M.C., Bijsterbosch, M.K., and van Berkel, T.J. 1992. Lactoferrin uptake by the rat liver. Characterization of the recognition site and effect of selective modification of arginine residues. *J. Biol. Chem.* 267:11229-11235.
375. Zimecki, M., Mazurier, J., Machnicki, M., Wieczorek, Z., Montreuil, J., and Spik.G. 1991. Immunostimulatory activity of lactotransferrin and maturation of CD4- and CD8- murine thymocytes. *Immunol. Lett.* 30:119-124.
376. Zimecki, M., and Machnicki, M. 1994. Lactoferrin inhibits the effector phase of the delayed type hypersensitivity to sheep erythrocytes and inflammatory reactions to *M. bovis* (BCG). *Arch. Immunol. Ther. Exp. (Warsz).* 42:171-177.
377. Zimecki, M., Miedzybrodzki, R., and Szymaniec, S. 1998a. Oral treatment of rats with bovine lactoferrin inhibits carrageenan-induced inflammation; correlation with decreased cytokine production. *Arch. Immunol. Ther. Exp. (Warsz)* 46:361-365.
378. Zimecki, M., Wlaszczyk, A., Cheneau, P., Brunel, A.S., Mazurier, J., Spik, G., and Kubler, A. 1998b. Immunoregulatory effects of a nutritional preparation containing bovine lactoferrin taken orally by healthy individuals. *Arch. Immunol. Ther. Exp. (Warsz)* 46:231-240.
379. Zimecki, M., Wlaszczyk, A., Zagulski, T., and Kubler, A. 1998c. Lactoferrin lowers serum interleukin 6 and tumor necrosis factor alpha levels in mice subjected to surgery. *Arch. Immunol. Ther. Exp. (Warsz)* 46:97-104.